# THE SCIENCE OF AIR

# THE
# SCIENCE
# OF
# AIR

## CONCEPTS & APPLICATIONS

## FRANK R. SPELLMAN, Ph.D.

TECHNOMIC
PUBLISHING CO., INC.
LANCASTER · BASEL

The Science of Air
aTECHNOMIC publication

Technomic Publishing Company, Inc.
851 New Holland Avenue, Box 3535
Lancaster, Pennsylvania 17604 U.S.A.

Printed in the United States of America
10   9   8   7   6   5   4   3   2   1

Main entry under title:
   The Science of Air: Concepts and Applications

A Technomic Publishing Company book
Bibliography: p.
Includes index p. 247

Library of Congress Catalog Card No. 98-88836
ISBN No. 1-56676-715-6

*To Loma Marie Spellman*

# Table of Contents

# Preface

AS a companion text to *The Science of Water: Concepts and Applications*, *The Science of Air: Concepts and Applications* deals with every aspect of air.

Important to this discussion about air and its critical importance on earth is man—man and his use, misuse, and reuse of air. This text takes the view that, since air is the primary substance and sustenance of most life on earth, it is precious—too precious to abuse, pollute, and ignore. The common thread woven throughout the fabric of this text is air resource utilization and its protection.

Written primarily as an information source, this text is not limited in its potential for other uses. For example, while this work can be used by environmental scientists or science practitioners to provide insight into the substance they work hard to study, collect, and treat, it can just as easily provide important information for the policy maker who may be charged with making decisions concerning air resource utilization. Consequently, this book will serve a varied audience: students, lay personnel, regulators, technical experts, attorneys, business leaders, and concerned citizens.

Why have a text on the science of air? The study of air is a science, one closely related to and interrelated with other scientific disciplines—chemistry, mathematics, meteorology, and physics, to name a few. To solve the problems and understand the issues related to air, air practitioners need a broad base of scientific information from which to draw.

This text is designed to fill that gap in air information that has been out there (in academia as well as the real world) for some time. *The Science of Air* is presented in a straightforward, informative, thought-provoking manner in a user-friendly format. Like its companion text, *The Science of Water,* this text serves up the big picture while creating a launching point for further knowledge, insight, and exploration. This text does not have all the answers—no text does—but maybe the questions it raises throughout will entice someone, somewhere, to start looking for answers.

# PART I

# Fundamentals

# CHAPTER 1

# Introduction

*Whether we characterize it as a caress, a gentle breeze, or a warm wind, or as a blustery gale, tempest, typhoon, tornado, or hurricane, air is vital. Air can dry, cool, warm, ventilate, or irritate us.*

*Air is scientifically unique. The combination of common and rare gases we breathe has made life possible. We cannot imagine life without breathing—we must constantly quench our "thirst" for air.*

*Air sustains growth. It creates the subtle and blatant movements that provide us with changing weather patterns. Air gives us the blessing of communication. From our first cry to our dying breath, our voices travel on a current of air.*

*Pure air is odorless, colorless, and tasteless. We rarely stop to think about it unless it brings something to us as a reminder. It covers Earth completely; nothing can escape air's touch. Air is life—life and air are inseparable. We sometimes call air the breath of life—a fitting name. But polluted air can become the bane of all life, capable, in time, of destroying all life as we know it.*

*Whether it provides the fundamental source of power in a sailing vessel or pushes the blades of a windmill, a billowy cloud, a dust mote, or a feather; whether it lifts a bird soaring on thermals; whether it wafts to us the sweet fragrance of gardenia, lavender, lilac, or rose—or a seed to fertile ground; whether it sets water lapping against some distant shore, drives a gritty wind that sculpts mountains to sand, or hammers a horrendous fist that flattens whatever stands in its path—cities, forests, crops, and man—air is essential.*

*Our very existence depends on air, but we have created a paradox with our vital line to life. Why would we abuse something so vital, something we cannot live without? Why do we foul the very essence of our lives? Why do we insult our environment at a faster pace than we can understand and mitigate the consequences? Why? Because air is everywhere. We've always had enough.*

*Let us hope that we always will. Let us hope that we are not destroying the very air we breathe. Let us hope that technology will aid us in our efforts to retain the quality of air we need to survive.*

*We need air as it should be: unpolluted and in the perfect mixture of elements we were evolved to inhale. Air doesn't care about the greenhouse effect, ozone depletion, or global warming, but we should. The bottom line is that we must not forget that we exist by environmental consent, subject to change without notice.*

## 1.1  SETTING THE STAGE

W HAT is air? Most of us would have little difficulty in quickly answering this question by stating that air is the oxygen that we breathe—the substance that we need to sustain life. We might also state that air surrounds us and is virtually everywhere; air is that substance we feel against our faces and skin when the wind blows; air is that gas that we use to fill our automobile tires; air is necessary for combustion to take place; hot air lifts our balloons; air under pressure (pneumatic air) powers our tools and machines; air can be any temperature.

What is air? The environmental scientist/practitioner would answer this question differently than most of us. He or she would state that air is a gas—that it is actually a combination of gases. The scientist might also state that a gas is a state of matter that is distinguished from the solid and liquid states by very low density and viscosity, relatively great expansion and contraction with changes in pressure and temperature, the ability to diffuse readily, and the spontaneous tendency to become distributed throughout any container.

An engineer might refer to air as a fluid because engineers may deal with fluid mechanics—the study of the behavior of fluids (including air) at rest or in motion. Fluids may be either gases or liquids. You are probably familiar with the physical difference between gases and liquids, as exhibited by air and water, but for the study of fluid mechanics it is convenient to classify fluids by their compressibility. Gases are very readily compressible (you've heard of compressed air), while liquids are only slightly compressible.

Air is the mixture of gases that constitutes the Earth's atmosphere. The atmosphere is that thin veil or envelope of gases that surrounds Earth like the skin of an apple—very thin, but very vital. The approximate composition of dry air by volume at sea level is 78% nitrogen, 21% oxygen (necessary for life as we know it), 0.93% argon, and 0.03% carbon dioxide, together with very small amounts of numerous other constituents (see Table 1.1). The water vapor content is highly variable and depends on atmospheric conditions. Air is said to be pure when none of the minor constituents are present in sufficient concentration to be injurious to the health of human beings or

TABLE 1.1. Composition of Air/Earth's Atmosphere.

| Gas | Chemical Symbol | Volume (%) |
|---|---|---|
| Nitrogen | $N_2$ | 78.08 |
| Oxygen | $O_2$ | 20.94 |
| Carbon dioxide | $CO_2$ | 0.03 |
| Argon | Ar | 0.093 |
| Neon | Ne | 0.0018 |
| Helium | He | 0.0005 |
| Krypton | Kr | trace |
| Xenon | Xe | trace |
| Ozone | $O_3$ | 0.00006 |
| Hydrogen | $H_2$ | 0.00005 |

animals, to damage vegetation, or to cause loss of amenities (e.g., through the presence of dirt, dust, or odors or by diminished sunshine).

Where does air come from? Many scientists state that 4.6 billion years ago, a cloud of dust and gases forged the Earth and created a dense molten core enveloped in cosmic gases. This was the proto-atmosphere or proto-air, composed mainly of carbon dioxide, hydrogen, ammonia, and carbon monoxide, but it did not last long before it was stripped away by a tremendous outburst of charged particles from the sun. As the outer crust of Earth began to solidify, a new atmosphere began to form from the gases pouring out from gigantic hot springs and volcanoes. This created an atmosphere composed of carbon dioxide, nitrogen oxides, hydrogen, sulfur dioxide, and water vapor. As the Earth cooled, water vapor condensed into highly acidic rainfall, which collected to form oceans and lakes.

For much of Earth's early existence, only trace amounts of free oxygen were present. Then green plants evolved in the oceans and began to add oxygen to the atmosphere as a waste gas. As millennia passed, oxygen increased to its present 21% of the atmosphere.

How do we know for sure about the evolution of air on Earth? There is no guessing involved with the geological record. Consider, for example, geological formations dated to 2 billion years ago. In these early sediments is a clear and extensive band of "red bed" sediments—sands colored with oxidized (ferric) iron. Previously, ferrous formations had been laid down that showed no oxidation. But there is more evidence. We can look at 4.5 billion years ago, when carbon dioxide in the atmosphere was beginning to be lost in sediments. The vast amount of carbon deposited in limestone, oil, and coal indicates that carbon dioxide concentrations must once have been many times greater than today, which stands at only 0.03%. The first carbonated deposits appeared about 1.7 billion years ago and the first sulfate deposits about 1 billion years ago. The decreasing carbon dioxide was balanced by an increase in the nitrogen content of the air. The forms of respiration advanced from fermentation 4 billion years ago, to anaerobic photosynthesis 3 billion years ago, to aerobic photosynthesis 1.5 billion years ago. The aerobic respiration that is so familiar today only began to appear about 500 million years ago.

Fast-forward to the present. The atmosphere itself continues to evolve, but human activities—with their highly polluting effects—have now overtaken nature in determining the changes. And that is the overriding theme of this text: human beings and their effect on Earth's air.

Have you ever wondered where air goes when we expel it from our lungs, or if, when we do so, it is still air? When we use air to feed our fires, power our machines, weld or braze our metals, vacuum our floors, or spray our paints, insecticides, lubricants, etc., do we change the nature of air? These questions and their answers are important in this text because this work explains the science of air, yet we do not know as much as we need to know about air. As a case in point, consider this: Have you ever gone to the library and tried to find a text that deals exclusively and extensively with the science of air? Such texts are few; there is a huge information gap out there.

To start with, let's talk about breathing air in particular—the air we need to survive, the air that probably concerns us the most.

When the average person takes in a deep breath of air, he/she probably gives little thought to what he or she is doing, that is, breathing life-sustaining air. The situation could be different, however. For example, consider a young woman who is a firefighter. On occasion, she has to fight fires wearing a self-contained breathing apparatus (SCBA) to avoid breathing smoke and decreased oxygen levels created by the fire. The standard SCBA with a single bottle contains approximately 45 minutes of air (Class D Breathing Air, which is regular air with 21% oxygen and associated gases, nitrogen, etc.).

On this particular day, our firefighter responds to a fire where she and another firefighter are required to enter a burning building to look for and rescue any trapped victims. The firefighters don their SCBAs, activate their air supply, and enter the burning structure. Normally, 45 minutes of air is plenty to make a quick survey of a house's interior, especially when it is on fire.

After having searched the first floor without discovering any victims, the two firefighters climb the stairs to the second floor. The fire, which started in the kitchen, is spreading fast, and the smoke and toxic vapors are spreading even faster. The firefighters know that any person within this house without respiratory protection will not survive for long. The fire may not kill them, but the smoke and toxic vapors surely will. At the landing upstairs, the firefighters are on their knees, and the smoke and toxic vapors are intensifying by the second. But the firefighters are not worried; they have all the air they need strapped to their backs.

By the time they reach the hallway, visibility is zero, the heat is intense, and the toxic vapors and smoke are so thick you can literally feel them. All is well until flames find their way up the stairs and quickly spread down the carpeted hallway to the backs of the firefighters. They have 15 minutes of air left. The situation has instantly changed from one of rescuing victims to one of fleeing for their own lives. They have 12 minutes of air left.

Their only hope of escape is through the flames, but they are not too worried—they are well equipped with fire protective clothing, and they have nine minutes of air left.

Normally, nine minutes of air is a lot of air in most escape situations, but this was no normal situation. As the firefighters slid down the stairs to the first floor, their air supply registered two minutes. They had used more air in the last 30 seconds than they had in the previous 10 minutes. This excessive use of air should come as no surprise when you consider that with toxic vapors, smoke, and flame all around, there would be a tendency to become frightened and to breath hard and fast, drawing in and expelling copious amounts of air. Even with their training, our two firefighters were no different from you or me: they were scared, and they breathed hard until they breathed their last. They fell unconscious right in front of the doorway—just one more breath of air with its 21% oxygen and they would have escaped. The irony is that the fire, well beyond its flashover state, had all the air (with its accompanying oxygen supply) it needed to continue its deadly destruction.

If we cannot live without air—if air is so precious, so necessary for sustaining life—then two questions arise: (1) Why do we ignore air? (2) Why do we abuse it (pollute it)?

We ignore air for the same reasons we ignore water: because it is so common, so accessible (normally), and so unexceptional that we can ignore it. Why do we pollute air? We'll discuss many of the reasons later in this text. Table 1.2 reports the number of fatalities during air pollution events in a number of places.

Air is one of our essential resources, sustaining life as it stimulates and pleases the senses. Although air itself is invisible to the human eye, it makes possible such sights as beautiful rainbows, heart-thrilling sunsets and sunrises, and the Northern Lights. Air is capable of many other wondrous things: a cool, soothing breeze can carry thousands of scents, both pungent and subtle: salty ocean breezes, approaching rain, fragrances from blooming flowers, and others.

The "others" concern us here: the sulfurous gases from industrial processes; the stink of garbage, refuse, and trash—all part of man's throwaways; the toxic remnants from pesticides, herbicides, and all the other "-cides." We are surrounded by it, but we seldom think about it until it displeases us. Pollution causes the problem, and as stated previously, we will cover this topic in greater detail later in the text.

TABLE 1.2. **Mortality Reported as a Result of Air Pollution Events.**

| Location | Year | Deaths |
|---|---|---|
| Belguim | 1930 | 63 |
| Pennsylvania | 1948 | 17 |
| London | 1948 | 700–800 |
| London | 1952 | 4000 |
| London | 1956 | 1000 |
| London | 1957 | 700–800 |
| London | 1959 | 200–250 |
| London | 1962 | 700 |
| London | 1963 | 700 |
| New York | 1963 | 200–400 |

*Source:* U.S. Senate Staff Report, 1968.

In the opening it was stated that a gap in knowledge exists in dealing with the science of air. This text is designed to bridge the gap, to fill in this obvious and unsatisfactory gap in information about air, because, on this planet, air is life.

## 1.2  SCOPE OF TEXT

Science is any systematic field of study or body of knowledge that aims through experiment, observation, and deduction to produce reliable explanations of phenomena in the physical world. The intent of this book is to present the science of air in a logical, step-by-step, plain English, user-friendly format.

Chapter 2 discusses gaseous oxygen in air and the natural gaseous and suspended minute liquids or solid particulates. Chapter 3 presents a thorough discussion of air mathematics (the type of math used when working in air pollution control technology). Chapter 4 discusses the physics of air, and Chapter 5 discusses basic air chemistry, because to gain even a basic understanding of air and air pollution control technology, it is important to understand air's chemical makeup. Chapter 6 presents a comprehensive discussion of Earth's atmosphere. Chapter 7 involves information about moisture in the atmosphere. Chapter 8 presents information on precipitation, evapotranspiration, and photosynthesis. Chapter 9 deals with atmospheric motion. Chapter 10 is all about weather and climate. Chapter 11 presents information dealing with microclimates. Chapter 12 discusses climatic change: greenhouse effect, global warming, and the rising sea level. In Chapter 13 we discuss air quality, while Chapter 14 presents information on air quality management (air quality regulations). Chapter 15 discusses air pollution, and Chapter 16 deals with air pollution control technology. The final chapter, Chapter 17, discusses indoor air pollution.

This book is specifically written to serve as a reference text for an undergraduate course and also as a guide to prepare air pollution control technologists for practice in the real world. It also serves as a basic primer for those who are interested in gaining knowledge about air-related topics. The paradigm (model or prototype) used here is based on real-world experience and application. The knowledge gained from work performed in the world of environmental science is what this text is all about.

## 1.3 DEFINITIONS OF KEY TERMS

Every branch of science, including air science, has its own language for communication. To work at even the edge of air science and the science disciplines closely related to air science, you must acquire a familiarity with the vocabulary used in this text.

While it is helpful for technical publications to include a glossary of key terms at the end of the work, for this text, including many of these key definitions early in the text facilitates a more orderly, logical, step-by-step learning activity. Thus, in the following section many of the key terms are listed and defined. Other terms not defined here will be defined when used in the text.

### 1.3.1 DEFINITIONS

- *Absolute pressure*   the total pressure in a system, including both the pressure of a substance and the pressure of the atmosphere (about 14.7 psi at sea level).
- *Acid*   any substance that releases hydrogen ions ($H^+$) when mixed into water.
- *Acid precipitation*   rain, snow, or fog containing higher than normal levels of sulfuric or nitric acid, which may damage forests, aquatic ecosystems, and cultural landmarks.
- *Acid surge*   a period of short, intense acid deposition in lakes and streams as a result of the release (by rainfall or spring snow melt) of acids stored in soil or snow.
- *Acidic solution*   a solution that contains significant numbers of ($H^+$) ions.
- *Airborne toxins*   hazardous chemical pollutants that have been released into the atmosphere and are carried by air currents.
- *Albedo*   reflectivity, or the fraction of incident light reflected by a surface.

- *Arithmetic mean*   a measurement of average value, calculated by summing all terms and dividing by the number of terms.
- *Arithmetic scale*   a series of intervals (equally spaced marks or lines), usually made along the side or bottom of a graph, that represent the range of values of the data.
- *Atmosphere*   a 500-kilometer thick layer of colorless, odorless gases known as air that surrounds the Earth and is composed of nitrogen, oxygen, argon, carbon dioxide, and other gases in trace amounts.
- *Atom*   the smallest particle of an element that still retains the characteristics of that element.
- *Atomic number*   the number of protons in the nucleus of an atom.
- *Atomic weight*   the sum number of protons and neutrons in the nucleus of an atom.
- *Base*   any substance that releases hydroxyl ions ($OH^-$) when it dissociates in water.
- *Chemical bond*   the force that holds atoms together within molecules. A chemical bond is formed when a chemical reaction takes place. Two types of chemical bonds are ionic bonds and covalent bonds.
- *Chemical reaction*   a process that occurs when atoms of certain elements are brought together and combine to form molecules, or when molecules are broken down into individual atoms.
- *Climate*   the long-term weather pattern of a particular region.
- *Covalent bond*   a type of chemical bond in which electrons are shared.
- *Density*   the weight of a substance per unit of its volume, for example, pounds per cubic foot.
- *Dew point*   the temperature at which a sample of air becomes saturated, that is, has a relative humidity of 100%.
- *Element*   any of more than 100 fundamental substances that consist of atoms of only one kind. Elements constitute all matter.
- *Emission standards*   the maximum amount of a specific pollutant permitted to be legally discharged from a particular source in a given environment.
- *Emissivity*   the relative power of a surface to reradiate solar radiation back into space in the form of heat or long-wave infrared radiation.
- *Energy*   the ability to do work, to move matter from place to place, or to change matter from one form to another.
- *First Law of Thermodynamics*   a natural law that states that, during a physical or chemical change, energy is neither created nor destroyed, but it may be changed in form and moved from place to place.
- *Global warming*   the increase in global temperature predicted to arise from increased levels of carbon dioxide, methane, and other greenhouse gases in the atmosphere.

- *Greenhouse effect*   the prevention of the reradiation of heat waves to space by carbon dioxide, methane, and other gases in the atmosphere. The greenhouse effect makes possible the conditions that enable life to exist on Earth.
- *Ion*   an atom or radical in solution carrying an integral electrical charge, either positive (cation) or negative (anion).
- *Insolation*   the solar radiation received by the Earth and its atmosphere—incoming solar radiation.
- *Lapse rate*   the rate of temperature change with altitude. In the troposphere, the normal lapse rate is −3.5°F per 1000 ft.
- *Matter*   anything that exists in time, occupies space, and has mass.
- *Mesosphere*   a region of the atmosphere (based on temperature) between approximately 35 to 60 miles in altitude.
- *Meteorology*   the study of atmospheric phenomena.
- *Mixture*   two or more elements, compounds, or both, mixed together with no chemical reaction occurring.
- *Ozone*   the compound $O_3$. Found naturally in the atmosphere in the ozonosphere, it is also a constituent of photochemical smog.
- *pH*   a means of expressing hydrogen ion concentration in terms of the powers of 10; a measurement of how acidic or basic a substance is. The pH scale runs from 0 (most acidic) to 14 (most basic). The center of the range (7) indicates the substance is neutral.
- *Photochemical smog*   an atmospheric haze that occurs above industrial sites and urban areas resulting from a reaction (which takes place in the presence of sunlight) between pollutants produced by high temperature and a pressurized combustion process (such as the combustion of fuel in a motor vehicle). The primary component of smog is ozone.
- *Photosynthesis*   the process of using the sun's light energy by chlorophyll-containing plants to convert carbon dioxide ($CO_2$) and water ($H_2O$) into complex chemical bonds, forming simple carbohydrates such as glucose and fructose.
- *Pollutant*   a contaminant at a concentration high enough to endanger the environment.
- *Pressure*   the force pushing on a unit area. Normally, in air applications pressure is measured in atmospheres, Pascal (Pa), or pounds per square inch (psi).
- *Primary pollutants*   pollutants emitted directly into the atmosphere, where they exert an adverse influence on human health and the environment. The six primary pollutants are carbon dioxide, carbon monoxide, sulfur oxides, nitrogen oxides, hydrocarbons, and particulates. All but carbon dioxide are regulated in the United States.

- *Rayleigh scattering*   the preferential scattering of light by air molecules and particles that accounts for the blueness of the sky. The scattering is proportional to $1/\lambda 4$.
- *Radon*   a naturally occurring radioactive gas arising from the decay of uranium-238, which may be harmful to human health in high concentrations.
- *Rain shadow effect*   the phenomenon that occurs as a result of the movement of air masses over a mountain range. As an air mass rises to clear a mountain, the air cools and precipitation forms. Often, both the precipitation and the pollutant load carried by the air mass will be dropped on the windward side of the mountain. The air mass is by then devoid of most of its moisture; consequently, the lee side of the mountain receives little or no precipitation and is said to lie in the rain shadow of the mountain range.
- *Relative humidity*   the concentration of water vapor in the air, which is expressed as the percentage that its moisture content represents of the maximum amount that the air could contain at the same temperature and pressure; the higher the temperature, the more water vapor the air can hold.
- *Secondary pollutants*   pollutants formed from the interaction of primary pollutants with other primary pollutants or with atmospheric compounds such as water vapor.
- *Second Law of Thermodynamics*   a natural law that states that, with each change in form, some energy is degraded to a less useful form and given off to the surroundings, usually as low-quality heat.
- *Solute*   the substance dissolved in a solution.
- *Solution*   a liquid containing a dissolved substance.
- *Specific gravity*   the ratio of the density of a substance to a standard density. For gases, the density is compared with the density of air, which equals 1.0.
- *Stratosphere*   an atmospheric layer extending from 6–7 miles to 30 miles above the Earth's surface.
- *Stratospheric ozone depletion*   the thinning of the ozone layer in the stratosphere, which occurs when certain chemicals (such as chlorofluorocarbons) capable of destroying ozone accumulate in the upper atmosphere.
- *Thermosphere*   an atmospheric layer that extends from 56 miles to outer space.
- *Troposphere*   the atmospheric layer that extends from the Earth's surface to 6–7 miles above the surface.
- *Weather*   the day-to-day pattern of precipitation, temperature, wind, barometric pressure, and humidity.
- *Wind*   horizontal air motion.

## 1.4 SUMMARY

The air you breathe is scented by the environment around you: the cup of coffee you just poured, the perfume or aftershave from the person at the next desk, a lingering dust of chalk in the air, secondhand cigarette smoke, the fumes from the traffic on the road, or the emissions from the industries in the area you live.

Our breathing is such an automatic reflex that we have no need to consciously breathe—if we stop, our bodies take over and breathe for us. Humans cannot go without breathing beyond a certain point (different and longer for people who have strengthened their ability) without our bodies taking control. The air we draw into our lungs with every breath is the one substance we must have to live—we are totally reliant on it minute-to-minute or else we die. Human beings can survive for weeks without food and for a few days without water, but without air, we succumb in a matter of minutes.

## 1.5 REFERENCE

*Air Quality Criteria*, staff report, Subcommittee on Air and Water Pollution, Committee on Public Works, U.S. Senate, pp. 94–411, July 1968.

# CHAPTER 2

# Air and Its Functions

*When informed that air is a predominately mechanical mixture of a variety of individual gases (most of which will not directly sustain life) forming the Earth's enveloping atmosphere, many students really don't give this information much thought. As with water and soil, air is just one of those substances that we all know we need, but that is as far as our thoughts generally go on the topic.*

*Our early ancestors may have viewed air in a different light; the historical record seems to support this. For even though they were ignorant about the nuts and bolts of science and nature, they understood the import of the sun, water, soil, and air. Evidence indicates that they paid homage to gods for each of these life-sustaining substances because they, unlike most of us, understood that these substances were a blessing and were to be held in the highest regard. Over time, we have come to recognize, to a degree, what our ancestors knew with surety from the very start: that the sun, water, soil, and air are fundamental to everything we know—to everything we will ever know. One fact is absolutely certain: if you cannot find air to breathe, the other three vital substances are of no importance.*

## 2.1 INTRODUCTION

AS implied in the chapter opening, air is something that we do not normally think about unless we have to: when we are short on air or when the air we breathe makes us cough, sick, or worse. When forced to think or talk about air, we generally don't think or talk about "air"—instead, we think or talk about the oxygen in the air. Gaseous oxygen sustains our lives via respiration, where the oxygen we take in "burns" the carbon in foodstuffs (like combustion), which then is exhaled. Other gaseous and particulate substances that make up air are not superfluous. While air and oxygen have become synonymous to most people, those of us who study or

**15**

practice science know that air and oxygen are very different, actually unique entities that just happen to blend or mix.

In this chapter, we talk about air—all about air—not just the gaseous oxygen in air, but the natural gaseous and suspended minute liquids or solid particulates (aerosols) that make up air. Even though this book is primarily directed toward a discussion about atmospheric air and the environmental implications affecting it, this chapter also addresses other important functions air provides: combustion and pneumatic power.

## 2.2 "REVOLUTIONARY" SCIENCE

Today, many people can explain in basic terms the composition of air, and they understand that the air we breathe contains oxygen, nitrogen, and other gases. Just a few hundred years ago, the actual composition of air was nothing more than speculation—a mystery.

The French aristocrat Antoine Lavoisier (1743–1794) is universally regarded as the founder of modern chemistry. This lofty title was bestowed on Lavoisier for his great experiments and discoveries related to the major components that make up air (oxygen and nitrogen) and to a lesser degree for identifying the components of water (hydrogen and oxygen).

Most of Lavoisier's experiments and discoveries took place in the years just preceding the French Revolution. Even though he ranks as one of the great scientists of his time, Lavoisier was guillotined on trumped-up charges during the French Revolution. Joseph Lagrange (1736–1813), the great French mathematician, said: "It required only a moment to sever his head, and probably 100 years will not suffice to produce another like it."

Lagrange's eulogy concerning Lavoisier and his scientific accomplishments is quite fitting. What seems simple and elementary to us today was not so clear 200 years ago. Indeed, the discoveries made at that time were quite difficult. We must remember that, in Lavoisier's time, so-called chemists had no clear idea of what a chemical element was nor any understanding of the nature of gases.

Lavoisier's discoveries were built on the work of others who preceded him or who were working on similar experiments during his lifetime. Lavoisier's work also provided a foundation for scientific discoveries that followed. For example, Lavoisier experimented with the findings of the German chemist Georg Stahl (1660–1743) and disproved them. Stahl proposed a theory that a combustible material burned because it contained a substance called *phlogiston* (charcoal is a prime example). Stahl knew that metallurgists obtained some metals from their ores by heating them with charcoal, which seemed to support the phlogiston theory of combustion.

However, in 1774 Lavoisier, with the help of Joseph Priestley (1733–1804), proved that the phlogiston theory was wrong. Priestley had heated a clax (in this particular case, the burned residue of oxide of mercury) in a closed apparatus and collected the gas liberated in the process. Priestley had discovered that this gas supported combustion better than air.

Lavoisier repeated Priestley's experiments and convinced himself of the presence in air of a gas that combined with substances when they burn and that it was the same gas given off when the oxide of mercury was heated. Thus, he proved that, when a substance burned, it combined with the oxygen in the air. He named this gas oxygine ("acid former" from the Greek) because he believed all acids contained oxygen.

In the meantime, Lavoisier had identified the other main component of air: nitrogen, which he named azote (from the Greek for "no life"). He also demonstrated that, when hydrogen (which chemists of the day called "inflammable air") was burned with oxygen, water was formed.

Lavoisier restructured chemistry and gave it its modern form. His work provided a firm foundation for the atomic theory proposed by British chemist and physicist John Dalton, and his elements were later classified in the periodic table. Lavoisier's work set the stage for the discovery of the other gaseous constituents in air, made later by other scientists.

With nitrogen and oxygen already identified as the primary constituents in air, and later carbon dioxide, water vapor, helium, ozone, and particulate matter, it was some time before the other gaseous constituents were identified. Argon was discovered in 1894 by British chemists John Rayleigh and William Ramsay after all oxygen and nitrogen had been chemically removed from a sample of air. Ramsay, along with Englishman Morris Travers, discovered neon; they also discovered krypton and xenon in 1889.

## 2.3 THE COMPONENTS OF AIR: CHARACTERISTICS AND PROPERTIES

It was pointed out that air is a combination of component parts: gases (see Table 1.1) and other matter (suspended minute liquid or particulate matter). In this section each of these components is discussed.

*Note:* Much of the information pertaining to atmospheric gases that follows was adapted from the Compressed Gas Association's *Handbook of Compressed Gases* (1990).

### 2.3.1 ATMOSPHERIC NITROGEN

Nitrogen ($N_2$) makes up the major portion of the atmosphere (78.03% by volume, 75.5% by weight). A colorless, odorless, tasteless, nontoxic, and almost totally inert gas, nitrogen is nonflammable, will not support combustion, and is not life-supporting. If gaseous nitrogen does not support life, what is it doing in our atmosphere? Nitrogen is indeed good, and without it, we could not survive.

Nitrogen is part of the Earth's atmosphere primarily because, over time, it has simply accumulated in the atmosphere and remained in place and in balance. This nitrogen accumulation process has occurred because, chemically, nitrogen is not very reactive. When released by any process, it tends not to recombine with other elements and accumulates in the atmosphere. This is good because we need nitrogen, not for breathing, but for other life-sustaining processes.

Let's take a look at a couple of reasons why gaseous nitrogen is so important to us. Although nitrogen in its gaseous form is of little use to us, after oxygen, carbon, and hydrogen, it is the most common element in living tissues. As a chief constituent of chlorophyll and amino acids and nucleic acids—the "building blocks" of proteins (which are used as structural components in cells)—nitrogen is essential to life. Animals cannot use nitrogen directly, but only when they obtain it by eating plant or animal tissues. Plants obtain the nitrogen they need in the form of inorganic compounds, principally nitrate and ammonium.

Gaseous nitrogen is converted to a form usable by plants (nitrate ions) chiefly through the process of nitrogen fixation via the nitrogen cycle, shown in simplified form in Figure 2.1.

Via the *nitrogen cycle,* aerial nitrogen is converted into nitrates mainly by microorganisms, bacteria, and blue-green algae. Lightning also converts some aerial nitrogen gas into forms that return to the Earth as nitrate ions in rainfall and other types of precipitation. From Figure 2.1 we see that ammonia plays a major role in the nitrogen cycle. Excretion by animals and anaerobic decomposition of dead organic matter by bacteria produce ammonia, which in turn is converted by nitrification bacteria into nitrites and then into nitrates. This process is known as nitrification. Nitrification bacteria are aerobic. Bacteria that convert ammonia into nitrites are known as nitrite bacteria (*Nitrosococcus* and *Nitrosomonas*). Although nitrite is toxic to many plants, it usually does not accumulate in the soil. Instead, other bacteria (such as *Nitrobacter*) oxidize the nitrite to form nitrate ($NO_3^-$), the most common biologically usable form of nitrogen.

Nitrogen reenters the atmosphere through the action of denitrifying bacteria, which are found in nutrient-rich habitats such as marshes and swamps. These bacteria break down nitrates into nitrogen gas and nitrous oxide ($N_2O$), which then reenter the atmosphere. Nitrogen also reenters the atmosphere from exposed nitrate deposits, from emissions from electric power plants and automobiles, and from volcanoes.

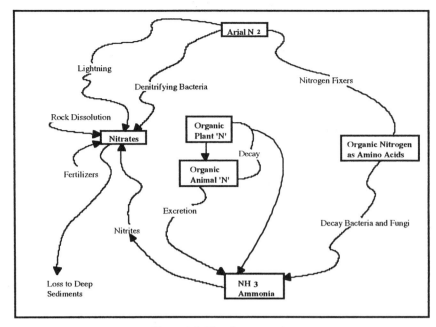

**Figure 2.1** The nitrogen cycle.

### 2.3.1.1 Nitrogen: Physical Properties

The physical properties of nitrogen are noted in Table 2.1.

### 2.3.1.2 Nitrogen: Uses

In addition to being the preeminent (in regards to volume) component of

TABLE 2.1. **Nitrogen: Physical Properties.**

| | |
|---|---|
| Chemical formula | $N_2$ |
| Molecular weight | 28.01 |
| Density of gas @ 70°F | 0.072 lb/ft$^3$ |
| Specific gravity of gas @ 70°F and 1 atm (air = 1) | 0.967 |
| Specific volume of gas @ 70°F and 1 atm | 13.89 ft$^3$/lb |
| Boiling point @ 1 atm | −320.4°F |
| Melting point @ 1 atm | −345.8°F |
| Critical temperature | −232.4°F |
| Critical pressure | 493 psia |
| Critical density | 19.60 lb/ft$^3$ |
| Latent heat of vaporization @ boiling point | 85.6 Btu/lb |
| Latent heat of fusion @ melting point | 11.1 Btu/lb |

Earth's atmosphere and providing an essential ingredient in sustaining life, nitrogen gas has many commercial and technical applications. As a gas, it is used in heat-treating of primary metals; the production of semiconductor electronic components (as a blanketing atmosphere); blanketing of oxygen-sensitive liquids and volatile liquid chemicals; inhibition of aerobic bacteria growth; and the propulsion of liquids through canisters, cylinders, and pipelines.

### 2.3.1.3 Nitrogen Oxides

There are six oxides of nitrogen: nitrous oxide ($N_2O$), nitric oxide (NO), dinitrogen trioxide ($N_2O_3$), nitrogen dioxide ($NO_2$), dinitrogen tetroxide ($N_2O_4$), and dinitrogen pentoxide ($N_2O_5$).

Nitric oxide, nitrogen dioxide, and nitrogen tetroxide are fire gases. One or more of them is generated when certain nitrogenous organic compounds (polyurethanes) burn. Nitric oxide is the product of incomplete combustion, and a mixture of nitrogen dioxide and nitrogen tetroxide is the product of complete combustion.

The nitrogen oxides are usually collectively symbolized by the formula $NO_x$. The USEPA, under the Clean Air Act (CAA), regulates the amount of nitrogen oxides that commercial and industrial facilities may emit to the atmosphere. The primary and secondary standards are the same: the annual concentration of nitrogen dioxide may not exceed 100 mg/m$^3$ (0.05 ppm).

Much more will be said about primary and secondary air standards under CAA and nitrogen oxides later in the book.

### 2.3.2 ATMOSPHERIC OXYGEN

Oxygen ($O_2$—Greek *oxys* "acid" *genes* "forming") constitutes approximately a fifth (21% by volume and 23.2% by weight) of the air in Earth's atmosphere. Gaseous oxygen ($O_2$) is vital to life as we know it. On Earth, oxygen is the most abundant element. Most oxygen on Earth is not found in the free state, but in combination with other elements as chemical compounds. Water and carbon dioxide are common examples of compounds that contain oxygen, but there are countless others.

At ordinary temperatures, oxygen is a colorless, odorless, tasteless gas that not only supports life, but also combustion. All the elements, except the inert gases, combine directly with oxygen to form oxides. However, oxidation of different elements occurs over a wide range of temperatures.

Oxygen is nonflammable, but it readily supports combustion. All materials that are flammable in air burn much more vigorously in oxygen. Some combustibles (such as oil and grease) burn with nearly explosive violence in oxygen if ignited.

### 2.3.2.1 Oxygen: Physical Properties

The physical properties of oxygen are noted in Table 2.2.

### 2.3.2.2 Oxygen: Uses

The major uses of oxygen stem from its life-sustaining and combustion-supporting properties. It also has many industrial applications (when used with other fuel gases such as acetylene), including metal cutting, welding, hardening, and scarfing.

### 2.3.2.3 Ozone: Just Another Form of Oxygen

*Ozone* ($O_3$) is a highly reactive pale-blue gas with a penetrating odor. Ozone is an allotropic modification of oxygen. An *allotrope* is a variation of an element that possesses a set of physical and chemical properties significantly different from the normal form of the element. Only a few elements have allotropic forms: oxygen, phosphorous, and sulfur are a few of them. Ozone is another form of oxygen. Formed when the molecule of the stable form of oxygen ($O_2$) is split by ultraviolet (UV) radiation or electrical discharge, it has three, instead of two, atoms of oxygen per molecule. Thus, its chemical formula is represented by $O_3$.

Ozone forms a thin layer in the upper atmosphere that protects life on Earth from ultraviolet rays (a cause of skin cancer). At lower atmospheric levels, it is an air pollutant and contributes to the greenhouse effect. At ground level, ozone, when inhaled, can cause asthma attacks, stunted growth in plants, and corrosion of certain materials. Produced by the action of sunlight on air pollutants (including car exhaust fumes), ozone is a major air pollutant in hot summers. We'll discuss ozone and the greenhouse effect more fully later in the text.

TABLE 2.2. **Oxygen: Physical Properties.**

| | |
|---|---|
| Chemical formula | $O_2$ |
| Molecular weight | 31.9988 |
| Freezing point | −361.12°F |
| Boiling point | −297.33°F |
| Heat of fusion | 5.95 Btu/lb |
| Heat of vaporization | 91.70 Btu/lb |
| Density of gas @ boiling point | 0.268 lb/ft$^3$ |
| Density of gas @ room temperature | 0.081 lb/ft$^3$ |
| Vapor density (air = 1) | 1.105 |
| Liquid-to-gas expansion ratio | 875 |

### 2.3.4 ATMOSPHERIC CARBON DIOXIDE

Carbon dioxide ($CO_2$) is a colorless, odorless gas (although some people feel it has a slight pungent odor and biting taste), slightly soluble in water, more dense than air (one and a half times heavier than air), and slightly acidic. Carbon dioxide gas is relatively nonreactive and nontoxic. It will not burn, and it will not support combustion or life.

$CO_2$ is normally present in atmospheric air at about 0.035% by volume and cycles through the biosphere (carbon cycle), as shown in Figure 2.2. Carbon dioxide, along with water vapor, is primarily responsible for the absorption of infrared energy re-emitted by the Earth. In turn, some of this energy is reradiated back to the Earth's surface. It is also a normal end product of human and animal metabolism. Our exhaled breath contains up to 5.6% carbon dioxide. Burning carbon-laden fossil fuels also releases carbon dioxide into the atmosphere. Much of this carbon dioxide is absorbed by ocean water, some of it is taken up by vegetation through photosynthesis in the carbon cycle (see Figure 2.2), and some remains in the atmosphere. Today, scientists estimate that the concentration of carbon dioxide in the atmosphere is approximately 350 parts per million (ppm) and rises at a rate of approximately 20 ppm every decade. The increasing rate of combustion of coal and oil has been primarily responsible for this occurrence, which (as we will see later in the book) may eventually have an impact on global climate.

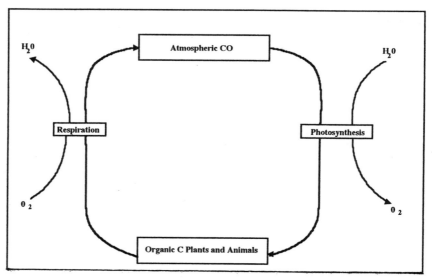

**Figure 2.2** Carbon cycle.

TABLE 2.3.  Carbon Dioxide: Physical Properties.

| | |
|---|---|
| Chemical formula | $CO_2$ |
| Molecular weight | 44.01 |
| Vapor pressure @ 70°F | 838 psig |
| Density of gas @ 70°F and 1 atm | 0.1144 lb/ft$^3$ |
| Specific gravity of gas @ 70°F and 1 atm (air = 1) | 1.522 |
| Specific volume of gas @ 70°F and 1 atm | 8.741 ft$^3$/lb |
| Critical temperature | −109.3°F |
| Critical pressure | 1070.6 psia |
| Critical density | 29.2 lb/ft$^3$ |
| Latent heat of vaporization @ boiling point | 100.8 Btu/lb |
| Latent heat of fusion @ −69.9°F | 85.6 Btu/lb |

## 2.3.4.1  Carbon Dioxide: Physical Properties

The physical properties of carbon dioxide are noted in Table 2.3.

## 2.3.4.2  Carbon Dioxide: Uses

Solid carbon dioxide is used extensively to refrigerate perishable foods while in transit. It is also used as a cooling agent in many industrial processes such as grinding, rubber work, cold-treating metals, vacuum cold traps, and so on.

Gaseous carbon dioxide is used to carbonate soft drinks, for pH control in water treatment, in chemical processing, as a food preservative, and in pneumatic devices.

## 2.3.5 ATMOSPHERIC ARGON

Argon (Ar—Greek *argos* "idle") is a colorless, odorless, tasteless, non-toxic, nonflammable gaseous element (noble gas). It constitutes almost 1% of the Earth's atmosphere and is plentiful compared to the other rare atmospheric gases. It is extremely inert and forms no known chemical compounds. It is slightly soluble in water.

### 2.3.5.1 Argon: Physical Properties

The physical properties of argon are noted in Table 2.4.

### 2.3.5.2 Argon: Uses

Argon is used extensively in filling incandescent and fluorescent light bulbs and electronic tubes; as a protective shield for growing silicon and

TABLE 2.4. **Argon: Physical Properties.**

| | |
|---|---|
| Chemical formula | Ar |
| Molecular weight | 39.95 |
| Density of gas @ 70°F | 0.103 lb/ft$^3$ |
| Specific gravity of gas @ 70°F | 1.38 |
| Specific volume of gas @ 70°F | 9.71 ft$^3$/lb |
| Boiling point @ 1 atm | −302.6°F |
| Melting point @ 1 atm | −308.6°F |
| Critical temperature | −188.1°F |
| Critical pressure | 711.5 psia |
| Critical density | 33.444 lb/ft$^3$ |
| Latent heat of vaporization @ boiling point and 1 atm | 69.8 Btu/lb |
| Latent heat of fusion | 12.8 Btu/lb |

germanium crystals; and as a blanket in the production of titanium, zirconium, and other reactive metals.

### 2.3.6 ATMOSPHERIC NEON

Neon (Ne—Greek *neon* "new") is a colorless, odorless, gaseous, nontoxic, chemically inert element. Air is about two parts per thousand neon by volume.

#### 2.3.6.1 Neon: Physical Properties

The physical properties of neon are noted in Table 2.5.

#### 2.3.6.2 Neon: Uses

Neon is used principally to fill lamp bulbs and tubes. The electronics

TABLE 2.5. **Neon: Physical Properties.**

| | |
|---|---|
| Chemical formula | Ne |
| Molecular weight | 20.183 |
| Density of gas @ 70°F and 1 atm | 0.05215 lb/ft$^3$ |
| Specific gravity of gas @ 70°F and 1 atm | 0.696 |
| Specific volume of gas @ 70°F and 1 atm | 9.18 ft$^3$/lb |
| Boiling point @ 1 atm | −410.9°F |
| Melting point @ 1 atm | −415.6°F |
| Critical temperature | −379.8°F |
| Critical pressure | 384.9 psia |
| Critical density | 30.15 lb/ft$^3$ |
| Latent heat of vaporization @ boiling point | 37.08 Btu/lb |
| Latent heat of fusion | 7.14 Btu/lb |

industry uses neon singly or in mixtures with other gases in many types of gas-filled electron tubes.

## 2.3.7 ATMOSPHERIC HELIUM

Helium (He—Greek *helios* "Sun") is inert (and as a result, does not appear to have any major effect on or role in the atmosphere), nontoxic, odorless, tasteless, nonreactive (forms no compounds), and colorless and constitutes about 0.00005% (5 ppm) by volume of the Earth's atmosphere. Helium, like neon, krypton, hydrogen, and xenon, is a noble gas. Helium is the second lightest element; only hydrogen is lighter. It is one-seventh as heavy as air. Helium is nonflammable and is only slightly soluble in water.

### 2.3.7.1 Helium: Physical Properties

The physical properties of helium are given in Table 2.6.

## 2.3.8 ATMOSPHERIC KRYPTON

Krypton (Kr—Greek *kryptos* "hidden") is a colorless, odorless, inert gaseous component of Earth's atmosphere. It is present in very small quantities in the air (about 114 ppm).

### 2.3.8.1 Krypton: Physical Properties

The physical properties of krypton are noted in Table 2.7.

### 2.3.8.2 Krypton: Uses

Krypton is used principally to fill lamp bulbs and tubes. The electronics industry uses it singly or in mixture in many types of gas-filled electron tubes.

TABLE 2.6. Helium: Physical Properties.

| | |
|---|---|
| Chemical formula | He |
| Molecular weight | 4.00 |
| Density of gas @ 70°F and 1 atm | 0.0103 lb/ft$^3$ |
| Specific gravity of gas @ 70°F and 1 atm | 0.138 |
| Specific volume of gas @ 70°F and 1 atm | 97.09 ft$^3$/lb |
| Boiling point @ 1 atm | −452.1°F |
| Critical temperature | −450.3°F |
| Critical pressure | 33.0 psia |
| Critical density | 4.347 lb/ft$^3$ |
| Latent heat of vaporization @ boiling point and 1 atm | 8.72 Btu/lb |

TABLE 2.7. **Krypton: Physical Properties.**

| | |
|---|---|
| Chemical formula | Kr |
| Molecular weight | 83.80 |
| Density of gas @ 70°F and 1 atm | 0.2172 lb/ft$^3$ |
| Specific gravity of gas @ 70°F and 1 atm | 2.899 |
| Specific volume of gas @ 70°F and 1 atm | 4.604 ft$^3$/lb |
| Boiling point @ 1 atm | −244.0°F |
| Melting point @ 1 atm | −251°F |
| Critical temperature | −82.8°F |
| Critical pressure | 798.0 psia |
| Critical density | 56.7 lb/ft$^3$ |
| Latent heat of vaporization @ boiling point | 46.2 Btu/lb |
| Latent heat of fusion | 8.41 Btu/lb |

## 2.3.9 ATMOSPHERIC XENON

Xenon (Xe—Greek *xenon* "stranger") is a colorless, odorless, nontoxic, inert, heavy gas that is present in very small quantities in the air (about 1 part in 20 million).

### 2.3.9.1 Xenon: Physical Properties

The physical properties of xenon are presented in Table 2.8.

### 2.3.9.2 Xenon: Uses

Xenon is used principally to fill lamp bulbs and tubes. The electronics industry uses it singly or in mixtures in many types of gas-filled electron tubes.

## 2.3.10 ATMOSPHERIC HYDROGEN

Hydrogen (H$_2$—Greek *hydros* + *gen* "water generator") is a colorless, odorless, tasteless, nontoxic, flammable gas. It is the lightest of all the elements and occurs on Earth chiefly in combination with oxygen as water. Hydrogen is the most abundant element in the universe, where it accounts for 93% of the total number of atoms and 76% of the total mass. It is the lightest gas known, with a density approximately 0.07 that of air. Hydrogen is present in the atmosphere, occurring in concentrations of only about 0.5 ppm by volume at lower altitudes.

TABLE 2.8. Xenon: Physical Properties.

| | |
|---|---|
| Chemical formula | Xe |
| Molecular weight | 131.3 |
| Density of gas @ 70°F and 1 atm | 0.3416 lb/ft$^3$ |
| Specific gravity of gas @ 70°F and 1 atm | 4.560 |
| Specific volume of gas @ 70°F and 1 atm | 2.927 ft$^3$/lb |
| Boiling point @ 1 atm | −162.6°F |
| Melting point @ 1 atm | −168°F |
| Critical temperature | 61.9°F |
| Critical pressure | 847.0 psia |
| Critical density | 68.67 lb/ft$^3$ |
| Latent heat of vaporization @ boiling point | 41.4 Btu/lb |
| Latent heat of fusion | 7.57 Btu/lb |

### 2.3.10.1 Hydrogen: Physical Properties

The physical properties of hydrogen are noted in Table 2.9.

### 2.3.10.2 Hydrogen: Uses

Hydrogen is used by refineries and petrochemical and bulk chemical facilities for hydrotreating, catalytic reforming, and hydrocracking. Hydrogen is used in the production of a wide variety of chemicals. Metallurgical companies use hydrogen in the production of their products. Glass manufacturers use hydrogen as a protective atmosphere in a process whereby molten glass is floated on a surface of molten tin. Food companies hydrogenate fats, oils, and fatty acids to control various physical and chemical

TABLE 2.9. Hydrogen: Physical Properties.

| | |
|---|---|
| Chemical formula | $H_2$ |
| Molecular weight | 2.016 |
| Density of gas @ 70°F and 1 atm | 0.00521 lb/ft$^3$ |
| Specific gravity of gas @ 70°F and 1 atm | 0.06960 |
| Specific volume of gas @ 70°F and 1 atm | 192.0 ft$^3$/lb |
| Boiling point @ 1 atm | −423.0°F |
| Melting point @ 1 atm | −434.55°F |
| Critical temperature | −399.93°F |
| Critical pressure | 190.8 psia |
| Critical density | 1.88 lb/ft$^3$ |
| Latent heat of vaporization @ boiling point | 191.7 Btu/lb |
| Latent heat of fusion | 24.97 Btu/lb |

properties. Electronic manufacturers use hydrogen at several steps in the complex processes for manufacturing semiconductors.

### 2.3.11 ATMOSPHERIC WATER

Leonardo da Vinci understood the importance of water when he said, "Water is the driver of nature." da Vinci was actually acknowledging what most scientists and many of the rest of us have come to realize: water, propelled by the varying temperatures and pressures in Earth's atmosphere, allows life as we know it to exist on our planet (Gradel & Crutzen, 1995).

The water vapor content of the lower atmosphere (troposphere) is normally in a range of 1–3% by volume, with a global average of about 1%. However, the percentage of water in the atmosphere can vary from as little as 0.1% to as much as 5%, depending upon altitude; water in the atmosphere decreases with increasing altitude. Water circulates in the atmosphere in the hydrologic cycle, as shown in Figure 2.3.

Water vapor contained in Earth's atmosphere plays several important roles: (1) it absorbs infrared radiation; (2) it acts as a blanket at night, retaining heat from the Earth's surface; and (3) it affects the formation of clouds in the atmosphere.

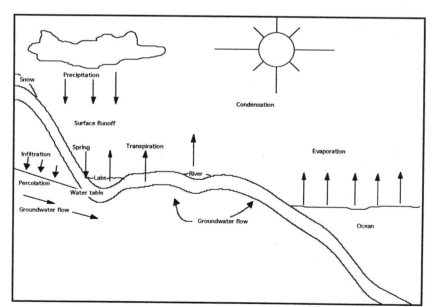

**Figure 2.3** Hydrologic cycle (water cycle). (*Source:* Adapted from Blackman, W. C., Jr., *Basic Hazardous Waste Management,* p. 53, 1993.)

## 2.3.12 ATMOSPHERIC PARTICULATE MATTER

Significant numbers of particles (particulate matter) are suspended in the atmosphere, particularly in the troposphere. These particles originate in nature from smokes, sea sprays, dusts, and the evaporation of organic materials from vegetation. A wide variety of nature's living or semi-living particles are also present—spores and pollen grains, mites and other tiny insects, and diatoms. The atmosphere also contains a bewildering variety of anthropogenic (man-made) particles produced by automobiles, refineries, production mills, and many other human activities.

Atmospheric particulate matter varies greatly in size. Colloidal-sized particles in the atmosphere are called *aerosols*—usually less than 0.1 mm in diameter; the smallest are gaseous clusters and ions and submicroscopic liquids and solids; somewhat larger ones produce the beautiful blue haze in distant vistas; those two to three times larger are highly effective in scattering light; and the largest consist of such things as rock fragments, salt crystals, and ashy residues from volcanoes, forest fires, or incinerators.

The concentration of particulates in the atmosphere varies greatly—ranging from more than 10,000,000/cubic centimeter to less than 1/L (0.001/cc). Excluding the particles in gases as well as vegetative material, sizes range from 0.005 to 500 microns, a variation in diameter of 100,000 times.

The largest number of airborne particulates is always in the invisible range. These numbers vary from less than one per liter to more than a half million per cubic centimeter in heavily polluted air and to at least ten times more than that when a gas-to-particle reaction is occurring (Schaefer & Day, 1981).

Based on particulate level, we can define two distinct regions in the atmosphere: very clean and dirty. The clean parts hold so few particulates that they are almost invisible, making them hard to collect or measure. In the dirty parts of the atmosphere—the air of a large metropolitan area—the concentration of particles includes an incredible quantity and variety of particulates from a wide variety of sources.

Atmospheric particulate matter performs a number of functions, undergoes several processes, and is involved in many chemical reactions in the atmosphere. Probably the most important function of particulate matter in the atmosphere is its action as nuclei for the formation of water droplets and ice crystals. Much of the work of Vincent J. Schaefer (inventor of cloud seeding) involved using dry ice in early attempts, but it later evolved around the addition of condensing particles to atmospheres supersaturated with water vapor and the use of silver iodide, which forms huge numbers of very small particles. Another important function of atmospheric particulate matter is that it helps determine the heat balance of the Earth's atmosphere by reflecting light. Particulate matter is also involved in many chemical reactions in the atmosphere—neutralization, catalytic effects, and oxidation reactions. These chemical reactions are discussed in greater detail in Chapter 4.

## 2.4  AIR FOR COMBUSTION

Try to imagine where the human race would be today or how far we would have progressed from our beginning to the present if we had not discovered and developed the use of fire. Today, of course, we are quite familiar with fire. We use the terms fire, combustion, oxidation, and burning interchangeably. However, there is a subtle difference between *combustion* and *oxidation*. During combustion, two or more substances chemically unite. In practice, one of them is almost always atmospheric oxygen, but combustion reactions are known in which oxygen is not one of the reactants. Thus, combustion is more correctly described as a *rapid oxidation*—or fire.

To state that atmospheric air plays an important role in combustion is to understate its significance; that is, we are stating what is readily apparent to anyone who has thought about what happens when someone snuffs a candle. Though air is important in combustion, the actual chemical reaction involved in combustion is something most of us give little thought to.

Combustion is a chemical reaction, one in which a fuel combines with air (oxygen) with the evolution of heat, i.e., burning. The combustion of fuels containing carbon and hydrogen is said to be *complete* when these two elements are oxidized to carbon dioxide and water (e.g., the combustion of carbon $C + O_2 = CO_2$). In air pollution control (Chapter 17), *incomplete combustion* concerns us because it may lead to (1) appreciable amounts of carbon remaining in the ash; (2) emission of some of the carbon as carbon monoxide; and (3) reaction of the fuel molecules to give a range of products that are emitted as smoke.

## 2.5  AIR FOR POWER

Along with performing its important function in Earth's atmosphere and its vital role in combustion, most industrial processes use gases to power systems of one type or another. The work is actually performed by a gas under pressure in the system. A gas power system may function as part of a process (such as heating and cooling), or it may be used as a secondary service system (such as compressed air). Compressed air is the gas most often found in industrial applications, but nitrogen and carbon dioxide are also commonly used. A system that uses a gas for transmitting force is called a *pneumatic* system (pneumatic is derived from the Greek word for an unseen gas). Originally, pneumatic referred only to the flow of air. Now it includes the flow of any gas in a system under pressure.

Pneumatic systems perform work in many ways, including operating air-driven tools, door openers, linear motion devices, and rotary motion devices. Have you ever watched (and heard) an automobile mechanic remove and replace a tire on your car? The device he uses to take off and put on tire lug nuts is a pneumatic wrench. Pneumatic hoisting equipment may be found in heavy fabricating environments, and pneumatic conveyors are used in the processing of raw materials. Pneumatic systems are also used to control flow valves in chemical process equipment and in large air-conditioning systems.

The pneumatic system in an industrial plant usually handles compressed air. As we said earlier, compressed air is used for operating portable air tools—drills, wrenches, and chipping tools for vises, chucks, and other clamping devices; for movable locating stops; for operating plastic molding machines; and also for supplying air used in manufacturing processes. Although the pieces of pneumatic equipment just described are different from each other, they all convert compressed air into work. Some of the laws of force and motion and their relation to pneumatic principles are reviewed in Chapter 5.

## 2.6 SUMMARY

We have gradually learned over the centuries how to make air a powerful tool. We harness its power, we separate it into its component parts, we compress it, we force it into useful configurations, and we use it to work quietly and cleanly for us. Air's makeup is a multipurpose array of essential ingredients, useful in many products—an array we are reliant on in more ways than we are readily aware of. The same air that sails a kite on a March wind can supply power to pneumatic tools, spin the sail of a windmill, and supply people with a breath of designer air in a chi-chi oxygen bar in San Francisco.

## 2.7 REFERENCES

*Compressed Gas Association, Inc., Handbook of Compressed Gases*, 3rd ed. New York: Van Nostrand Reinhold, 1990.

Graedel, T. E. & Crutzen, P. J., *Atmosphere, Climate, and Change.* New York: Scientific American, 1995.

Schaefer, V. J. & Day, J. A., *Atmosphere: Clouds, Rain, Snow, Storms.* Boston: Houghton Mifflin, 1981.

# CHAPTER 3

# Air Mathematics

*When studying a discipline that does not include mathematics, one thing is certain: the discipline under study has nothing or little to do with science.*

Put another way:

*If it can't be expressed in figures, it is not science; it is opinion. (Heinlein, 1973, p. 240)*

## 3.1 INTRODUCTION

**O**VER the years, countless studies have been conducted in an attempt to determine why many American students avoid science in school or avoid any branch of science as their chosen vocation. After years of study, many different opinions, and countless dollars spent on trying to determine why students avoid majoring in scientific disciplines, it is amazing that most of these studies did not go to the source to find the answers—the students, of course.

In having done just that, I found that most students avoid science or any thought of pursuing a vocation involving the sciences for two reasons: (1) many science curricula that lead to science-related vocations require completion of several hours of study in foreign languages, and (2) all scientific disciplines require an emphasis on mathematics. Many students I have spoken to cannot see the need to learn a foreign language. Some colleges require foreign language study with any four-year program; others do not.

Regarding the second requirement—a need for a strong background in mathematics to succeed in science—students have no way around or out of this necessity. To work with, around, or even at the periphery of science, you must have a strong background in mathematics. So why are students so fearful of math? Many of the students I asked said they dislike or fear math because it is difficult, time-consuming, and requires a lot of work. Let's take a look at how one math teacher puts it: "Those who have difficulty in math often do not lack the ability for mathematical calculation, they merely have not learned, or have been taught, the 'language of math' " (Price, 1991, p. vii). Price's point is well taken and can be expanded to: "The language of mathematics is a universal language." Mathematical symbols have the same meaning to people speaking in many different languages throughout the world.

What the problem with mathematics boils down to is that many students do not understand the language of mathematics. They are not familiar with the symbols, definitions, and terms of mathematics. You can't take shortcuts to learn this important subject—and some effort is required. Just as with any other subject, mathematics comes easily to some people and is difficult for others. Honest effort is all that is required to learn the language of mathematics.

What is mathematics? Mathematics is numbers. Math uses combinations of numbers and symbols to solve practical problems. Most of us, even the "mathophobes," handle what we have to in order to go through everyday life—we do the bills and balance the checkbook. Advanced scientific endeavors require a more serious exploration of higher mathematics, but for environmental science on this level, regular garden variety high school mathematics will fit the bill. Since we all use numbers every day, then we are all mathematicians, to a point.

In science, we must take math beyond "to a point." We need to learn, understand, and appreciate mathematics. But how do we do this without failing? Probably the greatest single cause of failure to understand and appreciate mathematics is not knowing the key definitions of the terms used. In mathematics, more than in any other subject, each word used has a definite and fixed meaning.

The following basic definitions should be memorized. They will aid you in learning the following material.

An *integer* (or an *integral number*) is a whole number. 1, 2, 3, 4, 5, 6, 7, 8, 9, 10, 11, and 12 are the first 12 positive integers.

A *factor* (or *divisor*) of a whole number is any other whole number that exactly divides it. Thus, 2 and 5 are factors of 10. A *prime number* in math is a number that has no factors except itself and 1. Examples of prime numbers are 1, 3, 5, 7, and 11.

A *composite number* is a number that has factors other than itself and 1. Examples of composite numbers are 4, 6, 8, 9, and 12.

A *common factor* (or *common divisor*) of two or more numbers is a factor that will exactly divide each of them. If this factor is the largest factor possible, it is called the *greatest common divisor*. Thus, 3 is a common divisor of 9 and 27, but 9 is the greatest common divisor of 9 and 27.

A *multiple* of a given number is a number that is exactly divisible by the given number. If a number is exactly divisible by two or more other numbers, that number is a common multiple of them. The least (smallest) such number is called the *lowest common multiple*. Thus 36 and 72 are common multiples of 12, 9, and 4; however, 36 is the lowest common multiple.

A *product* is the result of multiplying two or more numbers together. Thus, 25 is the product of $5 \times 5$.

A *quotient* is the result of dividing one number by another. For example, 5 is the quotient of 20 divided by 4.

A *dividend* is a number to be divided; a *divisor* is a number that divides. For example, in $100 \div 20 = 5$, 100 is the dividend, 20 is the divisor, and 5 is the quotient.

Mathematics has rules that we often have to apply in solving a mathematical problem. Sometimes the logic of the rule is apparent, but often it is not. Rules whose logic is not immediately obvious may be the result of experience, experiment, or merely "rules of thumb." We cannot solve mathematical problems without knowing the key definitions and without following the rules.

Since the study of air is a science, mathematics plays an important role in solving problems related to air problems. The mathematics review presented in the following sections is very basic. For those seeking advanced studies of air, higher mathematics is required. However, in this text, mathematics beyond basic algebra is not required.

## 3.2 UNITS OF MEASUREMENT

A fundamental knowledge of units of measurement and how to use them is essential for students of air science. Air science students and practitioners should be familiar with the U.S. Customary System (USCS) or English System and the International System of Units (SI). Some of the important units are summarized here to enable better understanding of material covered later in the text. Table 3.1 gives conversion factors between SI and USCS systems for some of the most basic units we will encounter.

In the study of air science, encountering both extremely large quantities and extremely small ones is very common. The concentration of some toxic substances may be measured in parts per million or billion (ppm or ppb). (For example, ppm may be roughly described as an amount contained in a shot glass in the bottom of a swimming pool.) To describe such large or small quantities, a system of prefixes that accompany the units is useful. Some of the more important prefixes are presented in Table 3.2.

### 3.2.1 UNITS OF MASS

Simply defined, *mass* is the quantity of matter in a material and the

TABLE 3.1. Commonly Used Units and Conversion Factors.

| Quantity | SI Units | SI Symbol × | Conversion Factor = | USCS Units |
|---|---|---|---|---|
| Length | meter | m | 3.2808 | ft |
| Mass | kilogram | kg | 2.2046 | lb |
| Temperature | Celsius | °C | 1.8 (°C) + 32 | °F |
| Area | square meter | $m^2$ | 10.7639 | $ft^2$ |
| Volume | cubic meter | $m^3$ | 35.3147 | $ft^3$ |
| Energy | kilojoule | kj | 0.9478 | Btu |
| Power | watt | W | 3.4121 | Btu/hr |
| Velocity | meter/second | m/s | 2.2369 | mi/hr |

measurement of the amount of inertia that it possesses. Mass expresses the degree to which an object resists a change in its state of rest or motion and is proportional to the amount of matter in the object.

Beginning science students often confuse mass with weight, but they are different. *Weight* is the gravitational force action upon an object and is proportional to mass. In the SI system (a modernized metric system), the fundamental unit of mass is the *gram (g)*. To show the relationship between mass and weight, consider that there are 452.6 grams per pound. In laboratory weighing operations, the gram is a convenient unit of measurement. However, in real-world applications, the gram is usually prefixed with one of the prefixes shown in Table 3.2. For example, human body mass is expressed in kilograms (1 kg = 2.2 pounds, which is the mass of one liter of water). When dealing with environmental conditions such as air pollutants and toxic water pollutants, they may be measured in teragrams ($1 \times 10^{12}$ grams) and micrograms ($1 \times 10^{-6}$ grams), respectively. When dealing with large-scale industrial commodities, the mass units may be measured in units of megagrams (Mg), which is also known as a metric ton.

TABLE 3.2. Common Prefixes.

| Quantity | Prefix | Symbol |
|---|---|---|
| $10^{-12}$ | pico | p |
| $10^{-9}$ | nano | n |
| $10^{-6}$ | micro | m |
| $10^{-3}$ | milli | m |
| $10^{-2}$ | centi | c |
| $10^{-1}$ | deci | d |
| 10 | deca | da |
| $10^2$ | hecto | h |
| $10^3$ | kilo | k |
| $10^6$ | mega | M |

Often, mass and density are mistakenly thought of as signifying the same thing; they do not. Where mass is the quantity of matter and measurement of the amount of inertia that a body contains, *density* refers to how compacted a substance is with matter. Density is the mass per unit volume of an object. Its formula can be written

$$\text{density} = \frac{\text{mass}}{\text{volume}} \tag{3.1}$$

Thus, something with a mass of 25 kg that occupies a volume of 5 m³ would have a density of 25 kg/5 m³ = 5 kg/m³.

### 3.2.2 UNITS OF LENGTH

In measuring locations and sizes, we use the fundamental property of *length,* which is defined as the measurement of space in any direction. Space has three dimensions, each of which can be measured by a length. This can be easily seen by considering the rectangular object shown in Figure 3.1. It has length, width, and height, but each of these dimensions is a length.

In the metric system, length is expressed in units based on the *meter* (m), which is 39.37 inches long. A kilometer (km) is equal to 1000 m, and is used to measure relatively great distances. In practical laboratory applications, the centimeter (cm = 0.01 m) is often used. There are 2.540 cm per inch, and the centimeter is employed to express lengths that would be given in inches in the English system. The micrometer (μm) is also commonly used to

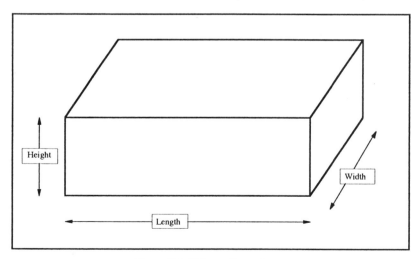

**Figure 3.1** Volume dimensions.

express measurements of bacterial cells and wavelengths of infrared radiation by which Earth reradiates solar energy back to outer space. For measuring visible light (400 to 800 nm), the nanometer (nm) ($10^{-9}$) is often used.

### 3.2.3 UNITS OF VOLUME

The easiest way in which to approach measurements involving volume is to remember that volume is surface area times a third dimension. The *liter* is the basic metric unit of volume and is the volume of a decimeter cubed (1 L = 1 $dm^3$). A milliliter (mL) is the same volume as a cubic centimeter ($cm^3$).

### 3.2.4 UNITS OF TEMPERATURE

*Temperature* is a measure of how "hot" something is or how much thermal energy it contains. Temperature is a fundamental measurement in air science, especially in most pollution work. The temperature of a stack gas plume, for example, determines its buoyancy and how far the plume of effluent will rise before attaining the temperature of its surroundings. This, in turn, determines how much it will be diluted before traces of the pollutant reach ground level.

Temperature is measured on several scales: for example, the *centigrade (Celsius)* and *Fahrenheit* scales are both measured from a reference point—the freezing point of water—which is taken as 0°C or 32°F. The boiling point of water is taken as 100°C or 212°F. For thermodynamic devices, it is usual to work in terms of absolute or thermodynamic temperature, where the reference point is absolute zero, which is the lowest possible temperature attainable. For absolute temperature measurement, the *Kelvin* (K) scale is used, which uses centigrade divisions for which zero is the lowest attainable measurement. A unit of temperature on this scale is equal to a Celsius degree, but is not called a degree; it is called a kelvin and is designated as K, not °K. The value of absolute zero on the Kelvin scale is −273.15°C, so that the Kelvin temperature is always a number 273 (rounded) higher than the Celsius temperature. Thus water boils at 373 K and freezes at 273 K.

It is easy to convert from the Celsius scale to the Kelvin scale. Simply add 273 to the Celsius temperature and you have the Kelvin temperature. Mathematically,

$$K = {}^\circ C + 273$$

where

K = temperature on the Kelvin scale
C = temperature on the Celsius scale

Converting from Fahrenheit to Celsius or vice versa is not as easy. The equations used are

$$C = 5/9\,(F - 32)$$

and

$$F = 9/5(C) + 32$$

where

C = temperature on the Celsius scale
F = temperature on the Fahrenheit scale

As examples, 15°C = 59°F and 68°F = 20°C. °F or °C, of course, can be negative numbers.

### 3.2.5  UNITS OF PRESSURE

*Pressure* is force per unit area and can be expressed in a number of different units, including the *atmosphere* (atm), which is the average pressure exerted by air at sea level, or the *pascal* (Pa), usually expressed in kilopascal (1 kPa = 1000 Pa and 101.3 kPa = 1 atm). Pressure can also be given as *millimeters of mercury* (mmHg), which is based on the pressure required to hold up a column of mercury in a mercury barometer. One mm of mercury is a unit called the *torr* and 760 torr equal 1 atm.

### 3.2.6  UNITS USED IN AIR STUDIES

In air studies, often the concentration of some substance (foreign or otherwise) in air is of interest. In either the gaseous or liquid medium, concentrations may be based on volume or weight or a combination of the two, which may lead to some confusion. To understand how weight and volume are used to determine concentrations when studying liquids or gases/vapors, the following explanations are provided.

#### 3.2.6.1  Liquids

Concentrations of substances dissolved in water are usually expressed in terms of weight of substance per unit volume of mixture. In environmental science, a good practical example of this weight per unit volume is best observed whenever a contaminant is dispersed in the atmosphere in solid or liquid form as a mist, dust, or fume. When this occurs, its concentration is usually expressed on a weight-per-volume basis. Outdoor air contaminants and stack effluents are frequently expressed as grams, milligrams, or micrograms per cubic meter; ounces per thousand cubic feet; pounds per thousand pounds of air; and grains per cubic foot. Most measurements are expressed in metric units. However, the use of standard U.S. units is justified for purposes of comparison with existing data, especially those relative to the specifications for air-moving equipment.

Alternatively, concentrations in liquids are expressed as weight of substance per weight of mixture, with the most common units being parts per million (ppm) or parts per billion (ppb). Since most of the concentrations of pollutants are very small, 1 L of mixture weighs essentially 1000 g, so we can write

$$1 \, mg / L = 1 \, g / m^3 = 1 \, ppm \, (by \, weight)$$

$$1 \, \mu g / L = 1 \, mg / m^3 = 1 \, ppb \, (by \, weight)$$

The air science practitioner may also be involved with concentrations of liquid wastes (which may contaminate the atmosphere) that are so high that the *specific gravity* (the ratio of an object's or substance's weight to that of an equal volume of water) of the mixture is affected, in which case a correction to the above may be required:

$$mg / L = ppm \, (by \, weight) \times specific \, gravity \quad (3.2)$$

### 3.2.6.2 Gases/Vapors

For most air pollution work, expressing pollutant concentrations in volumetric terms is customary. For example, the concentration of a gaseous pollutant in parts per million (ppm) is the volume of pollutant per million parts of the air mixture. That is,

$$ppm = \frac{parts \, of \, contamination}{million \, parts \, of \, air} \quad (3.3)$$

Note that calculations for gas and vapor concentrations are based on the gas laws:

- The volume of gas under constant temperature is inversely proportional to the pressure.
- The volume of a gas under constant pressure is directly proportional to

the Kelvin temperature. The Kelvin temperature scale is based on absolute zero (0°C = 273 K).
- The pressure of a gas of a constant volume is directly proportional to the Kelvin temperature.

Thus, when measuring contaminant concentrations, you must know the atmospheric temperature and pressure under which the samples were taken. At *standard temperature and pressure* (STP), 1 g-mol of an ideal gas occupies 22.4 L. The STP is 0°C and 760 mmhg. If the temperature is increased to 25°C (room temperature) and the pressure remains the same, 1 g-mol of gas occupies 24.45 L.

Sometimes it is necessary to convert milligrams per cubic meter ($mg/m^3$)—a weight-per-volume ratio—into a volume-per-unit weight ratio. If it is understood that 1 g-mole of an ideal gas at 25°C occupies 24.45 L, the following relationships can be calculated:

$$\text{ppm} = \frac{24.45}{\text{molecular wt}} \, mg/m^3 \tag{3.4}$$

$$mg/m^3 = \frac{\text{molecular wt}}{24.45} \, \text{ppm} \tag{3.5}$$

## 3.3 BASIC MATH

*Note*: It is assumed the reader has a fundamental knowledge of basic mathematical operations. Thus, the purpose of the following sections is to provide only a brief review of the mathematical concepts and applications frequently employed in actual air science activities.

## 3.4 POWERS OF TEN AND SCIENTIFIC NOTATION

*Note:* In practice, the air science practitioner should realize that the accuracy of a final answer can never be better than the accuracy of the data used. Furthermore, remember that correct and accurate data are worthless unless the operator is able to make correct computations.

We describe two common methods of expressing numbers in this section—powers of ten and scientific notation.

### 3.4.1 POWERS OF TEN NOTATION

An expression such as $5^7$ is a shorthand method of writing multiplication. For example, $5^7$ can be written as

$$5 \times 5 \times 5 \times 5 \times 5 \times 5 \times 5$$

The expression $5^7$ is referred to as 5 to the seventh power. It is composed of an exponent and a base number. The exponent (or power of) indicates how many times a number is to be multiplied together. The base is the number being multiplied.

$$5^{7} \leftarrow \text{(exponent)}$$
$$\nwarrow \text{(base)}$$

These same considerations apply to letters ($a$, $b$, $x$, $y$, etc.) as well. For example,

$$z^{2} = (z)(z) \text{ or } z^{4} = (z)(z)(z)(z)$$

When a number or letter does not have an exponent, it is considered to have an exponent of one.

$$\text{Thus } 5 = 5^{1} \text{ or } z = z^{1}$$

How is the term $(3/8)^2$ written in expanded form? When parentheses are used, the exponent refers to the entire term within the parentheses. Thus,

$$(3/8)^{2} = (3/8)(3/8)$$

When a negative exponent is used with a number or term, a number can be re-expressed using a positive exponent:

$$6^{-3} = 1/6^{3}$$

*Note:* Any number or letter such as $3^0$ or $X^0$ does not equal $3 \times 1$ or $X1$, but simply 1.

When a term is given in expanded form, you can determine how it would be written in exponential form. For example,

$$(5)(5)(5) = 5^{3}$$

or

$$(\text{in.})(\text{in.}) = \text{in.}^{2}$$

We commonly see powers used with a number or term used to denote area or volume units such as square inches or cubic feet ($\text{in.}^2$, $\text{ft}^2$, $\text{in.}^3$, $\text{ft}^3$).

When the exponents of both the numerator and denominator are the same, parentheses may be placed around the fraction using a single exponent outside of the parentheses, as follows:

$$\frac{(in.)(in.)}{(ft)(ft)} = \frac{(in.)^2}{(ft)^2} = \left(\frac{in.}{ft.}\right)^2$$

In moving a power from the numerator of a fraction to the denominator or vice versa, the sign of the exponent is changed. For example,

$$\frac{3^3 \times 4^{-2}}{8} = \frac{3^3}{8 \times 4^2}$$

## 3.4.2 SCIENTIFIC NOTATION

*Scientific notation* is a method by which any number can be expressed as a term multiplied by a power of ten. The term is always greater than or equal to 1 but less than 10. Examples of powers of ten are

$$3.2 \times 10^1$$

$$1.8 \times 10^3$$

$$9.550 \times 10^4$$

$$5.31 \times 10^{-2}$$

The numbers can be taken out of scientific notation by performing the indicated multiplication. For example,

$$3.2 \times 10^1 = (3.2)(10) = 32$$

$$1.8 \times 10^3 = (1.8)(10)(10)(10) = 1,800$$

$$9.550 \times 10^4 = (9.550)(10)(10)(10)(10) = 95,500$$

$$5.31 \times 10^{-2} = (5.31)1/10^2 = 0.0531$$

An easier way to take a number out of scientific notation is by moving the decimal point the number of places indicated by the exponent.

---

*RULE 1*

Multiply by the power of ten indicated. A positive exponent indicates a decimal move to the *right*, and a negative exponent indicates a decimal move to the *left*.

---

Consider the third example above:

$$9.550 \times 10^4$$

The positive exponent of 4 indicates that the decimal point should be moved four places to the right:

$$9.5500 = 95,000$$

The final example is

$$5.31 \times 10^{-2}$$

The negative exponent of 2 indicates that the decimal point should be moved two places to the left:

$$05.31 = 0.0531$$

There are very few instances in which you will need to put a number into scientific notation, but you should know how to do it, if required.

*Procedure*: When placing a number into scientific notation, place a decimal point after the first nonzero digit. (Remember, if no decimal point is shown in the number to be converted, it is assumed to be at the end of the number.) Count the number of places from the standard position to the original decimal point. This represents the exponent of the power of ten.

---

*RULE 2*

When a number is put into scientific notation, a decimal point move to the *left* indicates a *positive* exponent; a decimal point move to the *right* indicates a *negative* exponent.

---

Now let's try converting a few numbers into scientific notation. Let's try converting 1,500. Remember, to obtain a number between 1 and 9, the decimal point must be moved three places to the left. The number of place moves (3) becomes the exponent of the power of 10, and the move to the left indicates a positive exponent:

$$1,500 = 1.5 \times 10^3$$

Let's try a decimal number 0.0661:

$$0.0661 = 6.61 \times 10^{-2}$$

Two place moves to the right indicates a negative exponent of 2.

## 3.5 DIMENSIONAL ANALYSIS

*Dimensional analysis* is a valuable tool used as a way to check if you have set up a problem correctly. In using dimensional analysis to check a math setup, you work with the dimensions (units of measure) only—not with numbers. To use the dimensional analysis method, you must know how to perform three basic operations:

(1) To complete a division of units, always ensure that all units are written in the same format; it is best to express a horizontal fraction (such as gal/ft³) as a vertical fraction:

$$\text{gal / cu ft to } \frac{\text{gal}}{\text{cu ft}}$$

$$\text{psi to } \frac{\text{lb}}{\text{sq in.}}$$

(2) You must know how to divide by a fraction. The standard rule is to invert the terms in the denominator and then multiply the fractions. For example,

$$\frac{\dfrac{\text{lb}}{\text{d}}}{\dfrac{\text{min}}{\text{d}}} \text{ becomes } \frac{\text{lb}}{\text{d}} \times \frac{\text{d}}{\text{min}}$$

(3) You must know how to cancel or divide terms in the numerator and denominator of a fraction.

After fractions have been rewritten in the vertical form and division by the fraction has been re-expressed as multiplication as shown above, then the terms can be canceled (or divided) out. For every term that is canceled in the numerator of a fraction, a similar term must be canceled in the denominator and vice versa, as shown below:

$$\frac{kg}{\cancel{d}} \times \frac{\cancel{d}}{min} = \frac{kg}{min}$$

$$\cancel{mm^2} \times \frac{m^2}{\cancel{mm^2}} = m^2$$

$$\frac{\cancel{gal}}{min} \times \frac{ft^3}{\cancel{gal}} = \frac{ft^3}{min}$$

*Question*: How are units that include exponents calculated?

When written with exponents (such as $ft^3$), a unit can be left as is or put in expanded form, (ft)(ft)(ft), depending upon other units in the calculation. The point is that you must ensure that square and cubic terms are expressed uniformly, as sq ft and cu ft, or as $ft^2$ and $ft^3$. For dimensional analysis, the latter system is preferred.

For example, let's say that you wish to convert 1,400 $ft^3$ volume to gallons, and you will use 7.48 $gal/ft^3$ in the conversion. Do you multiply or divide by 7.48? You can use dimensional analysis to answer this question. To determine if the math setup is correct, only the dimensions are used.

First, try dividing the dimensions:

$$\frac{ft^3}{gal/ft^3} = \frac{ft^3}{\dfrac{gal}{ft^3}}$$

Then the numerator and denominator are multiplied to get

$$\frac{ft^6}{gal}$$

So, by dimensional analysis, you determine that, if you divide the two dimensions, the units of the answer are $ft^6/gal$, not gal. It is clear that division is not the right way to make this conversion.

What would have happened if you had multiplied the dimensions instead of dividing?

$$(ft^3)(gal/ft^3) = (ft^3)\left(\frac{gal}{ft^3}\right)$$

Then multiply the numerator and denominator to obtain

$$\left( \frac{(\text{ft}^3)(\text{gal})}{\text{ft}^3} \right)$$

and cancel common terms to obtain

$$\left( \frac{(\cancel{\text{ft}^3})(\text{gal})}{\cancel{\text{ft}^3}} \right)$$

$$= \text{gal}$$

Obviously, by multiplying the two dimensions, the answer will be in gallons, which is what you want. Thus, since the math setup is correct, you would then multiply the numbers to obtain the number of gallons.

$$(1{,}400 \text{ ft}^3)\,(7.48 \text{ gal}/\text{ft}^3) = 10{,}472 \text{ gal}$$

Now let's try another problem with exponents. You wish to obtain an answer in square feet, given the two terms 70 ft³/s and 4.5 ft/s. First, only the dimensions are used to determine if the math setup is correct. By multiplying the two dimensions, you get

$$(\text{ft}^3/\text{s})\,(\text{ft}/\text{s}) = \left( \frac{\text{ft}^3}{\text{s}} \right)\left( \frac{\text{ft}}{\text{s}} \right)$$

Then multiply the term in the numerators and denominators of the fraction:

$$\frac{(\text{ft}^3)\,(\text{ft})}{(\text{s})\,(\text{s})}$$

$$= \frac{\text{ft}^4}{\text{s}^2}$$

Obviously, the math setup is incorrect, because the dimensions of the answer are not square feet. Let's try division of the two dimensions instead.

$$\frac{\dfrac{\text{ft}^3}{\text{s}}}{\dfrac{\text{ft}}{\text{s}}}$$

Invert the denominator and multiply to get

$$\left(\frac{ft^3}{(s)}\right)\left(\frac{s}{(ft)}\right)$$

$$=\frac{(ft)\,(ft)\,(ft)\,(s)}{(s)\,(ft)}$$

$$=\frac{(\cancel{ft})\,(ft)\,(ft)\,(\cancel{s})}{(\cancel{s})\,(\cancel{ft})}$$

$$=ft^2$$

Since the dimensions of the answer are square feet, this math setup is correct. Therefore, by dividing the numbers, as was done with the units, the answer will also be correct.

$$\frac{70\,ft^3\,/\,s}{4.5\,ft\,/\,s}=15.56\,ft^2$$

## *Example 1*

You are given two terms, 5 m/s and 7 m², and the answer to be obtained is in cubic meters per second (m³/s). Is multiplying the two terms the correct math setup?

$$(m/s)\,(m^2)=\frac{m}{s}\times m^2$$

Multiply the numerators and denominator of the fraction:

$$\frac{(m)(m^2)}{s}$$

$$=\frac{m^3}{s}$$

Since the dimensions of the answer are cubic meters per second (m³/s), the math setup is correct. Therefore, multiply the numbers to get the correct answer.

$$(5\,m/s)(7\,m^2)=35\,m^3\,/\,s$$

*Example 2*

Solve the following problem: The flow rate in a water line is 2.3 ft³/s. What is the flow rate expressed as gallons per minute?

Set up the math problem and then use dimensional analysis to check the math setup:

$$(2.3 \text{ ft}^3 / \text{s})(7.48 \text{ gal} / \text{ft}^3)(60 \text{ s} / \text{min})$$

Dimensional analysis is used to check the math setup:

$$\left(\text{ft}^3 / \text{s}\right)\left(\text{gal} / \text{ft}^3\right)(\text{s} / \text{min}) = \left(\frac{\text{ft}^3}{\text{s}}\right)\left(\frac{\text{gal}}{\text{ft}^3}\right)\left(\frac{\text{s}}{\text{min}}\right)$$

$$= \left(\frac{\cancel{\text{ft}^3}}{\cancel{\text{s}}}\right)\left(\frac{\text{gal}}{\cancel{\text{ft}^3}}\right)\left(\frac{\cancel{\text{s}}}{\text{min}}\right)$$

$$= \frac{\text{gal}}{\text{min}}$$

The math setup is correct as shown above. Therefore, this problem can be multiplied out to get the answer in correct units.

$$(2.3 \text{ ft}^3 / \text{s})(7.48 \text{ gal} / \text{ft}^3)(60 \text{ s} / \text{min}) = 1032.24 \text{ gal} / \text{min}$$

## 3.6  ROUNDING OFF A NUMBER

Sometimes rounding off measurements to a certain number of significant figures is necessary. The number of significant figures in a measurement is the number of digits that are known for sure, plus one more digit that is an estimate. The number 6.93436 cm, for example, has six significant figures. For numbers less than one (such as 0.00696), the zeros to the right of the decimal point are not considered significant figures; thus, 0.00696 has three significant figures. For numbers greater than ten (such as 696,000), the zeros should not be considered significant; thus, the number 696,000 has three significant figures.

Rounding off a number means replacing the final digit of a number with zeros, thus expressing the number as tens, hundreds, thousands, ten thousands, or tenths, hundredths, thousandths, ten thousandths, etc. (e.g., 498 as 500; 0.49 as 0.5; or 6,696,696 as 7,000,000).

There is a basic rule to be followed when rounding off numbers.

---

### *RULE 3*

A number is rounded off by dropping one or more numbers from the right and adding zeros if necessary to place the decimal point. If the last figure dropped is five or more, increase the last retained figure by one. If the last figure dropped is less than five, do not increase the last retained figure.

---

***Example 3:*** Rounding to Significant Figures

Round off 11,548 to 4, 3, 2, and 1 significant figure.

*Solutions*

$$11,548 = 11,550 \text{ to 4 significant figures}$$

$$11,548 = 11,500 \text{ to 3 significant figures}$$

$$11,548 = 11,000 \text{ to 2 significant figures}$$

$$11,548 = 10,000 \text{ to 1 significant figure}$$

***Example 4:*** Rounding to a Particular Place Value in the Decimal System

Round 47,936 to the nearest hundred.

The procedure used in this rounding depends on the digit just to the right of the hundreds place:

$$47,936$$
$$\uparrow$$
(hundreds place)

Since the digit to the right of the hundreds place is less than 5, 9 is not changed, and all the digits to the right of 9 are replaced with zeros: 47,936 becomes 47,900 (rounded to the nearest hundred).

Let's look at another example where a decimal number is to be rounded. Round 6.654 to the nearest tenth.

$$6.654$$
$$\uparrow$$
(tenths place)

The digit to the right of the tenths place is 5; therefore, the 6 is increased by 1 and all digits to the right are dropped: 6.654 becomes 6.7 (rounded to the nearest tenth).

## 3.7 EQUATIONS: SOLVING FOR THE UNKNOWN

In environmental science applications related to air measurements and calculations, you may use equations to solve for the unknown quantity. To make these calculations, you must first know the values for all but one of the terms of the equation to be used.

An *equation* is a statement that two expressions or quantities are equal in value. The statement of equality $5x + 4 = 19$ is an equation; that is, it is algebraic shorthand for "The sum of 5 times a number plus 4 is equal to 19." It can be seen that the equation $5x + 4 = 19$ is much easier to work with than the equivalent sentence.

When thinking about equations, consider an equation as similar to a balance. The equals sign tells you that two quantities are "in balance" (i.e., they are equal).

Let's get back to the equation $5x + 4 = 19$. The solution to this problem may be summarized in three steps.

$$(1) \qquad 5x + 4 = 19$$

$$(2) \qquad 5x = 15$$

$$(3) \qquad x = 3$$

Step 1 expresses the whole equation. In Step 2, 4 has been subtracted from both members of the equation. In Step 3, both members have been divided by 5.

An equation is therefore kept in balance (both sides of the equation are kept equal) by subtracting the same number from both members (sides), adding the same number to both, or dividing or multiplying by the same number.

The expression $5x + 4 = 19$ is called a conditional equation because it is true only when $x$ has a certain value. The number to be found in a conditional equation is called the unknown number, the unknown quantity, or, more briefly, the unknown.

Solving an equation is finding the value or values of the unknown that make the equation true.

Another equation, one that air practitioners should be familiar with, is

$$W = F \times D \qquad\qquad (3.6)$$

where

$W$ = work
$F$ = force
$D$ = distance

Thus,

$$\text{work} = \text{force (pounds)} \times \text{distance (feet or inches)}$$

$$= \text{foot-pounds or inch-pounds}$$

To demonstrate another equation the air science practitioner may be called upon to use, consider the following situation. Fabric filters are commonly used to separate dry particles from a gas stream, usually of air or combustion gases. In fabric filtration, the particulate-laden gas flows into and through a number of filter bags placed in parallel, leaving the particulates retained by the fabric (more will be said about fabric filters in Chapter 17).

$$V = \frac{Q}{A} \tag{3.7}$$

where

$V$ = superficial filtering velocity (aka air / cloth ratio)
$Q$ = volumetric gas flow rate, $m^3$ / min
$A$ = cloth area, $m^3$

The terms of these equations are $W, F, D$ and $V, Q, A$. In solving problems using these equations, you would need to be given values to substitute for any two of the three terms in each equation. Again, the term for which you do not have information is called the unknown, which is often indicated by a letter such as $x, y,$ or $z$, but it may be any letter.

Suppose you have this equation:

$$80 = (x)(4)$$

How can you determine the value of $x$? By following the axioms presented below, the solution to the unknown is quite simple.

### 3.7.1 AXIOMS

(1) If equal numbers are added to equal numbers, the sums are equal.
(2) If equal numbers are subtracted from equal numbers, the remainders are equal.
(3) If equal numbers are multiplied by equal numbers, the products are equal.
(4) If equal numbers are divided by equal numbers (except zero), the quotients are equal.

(5) Numbers that are equal to the same number or to equal numbers are equal to each other.

(6) Like powers of equal numbers are equal.

(7) Like roots of equal numbers are equal.

(8) The whole of anything equals the sum of all of its parts.

*Note:* Axioms 2 and 4 were used to solve the equation $5x + 4 = 19$.

For example, find the value of $x$ if $x - 6 = 3$.

Here you can see by inspection that $x = 9$, but inspection does not help in solving more complicated equations. But if you notice that to determine that $x = 9$, 6 is added to each member of the given equation, you have acquired a method or procedure that can be applied to similar but more complex problems.

*Solution*: Given equation:

$$x - 6 = 3$$

Add 6 to each member (axiom 1),

$$x = 3 + 6$$

Collecting the terms (that is, adding 3 and 6),

$$x = 9$$

## 3.7.2 CHECKING THE ANSWER

After you have obtained a solution to an equation, you should always check it—an easy process. All you need to do is substitute the solution for the unknown quantity in the given equation. If the two members of the equation are then identical, the number substituted is the correct answer.

**Example 5**

Solve and check $4x + 5 - 7 = 2x + 6$

*Solution*

$$4x + 5 - 7 = 2x + 6$$
$$4x - 2 = 2x + 6$$
$$4x = 2x + 8$$

$$2x = 8$$

$$x = 4$$

Substituting the answer $x = 4$ in the original equation,

$$4x + 5 - 7 = 2x + 6$$

$$4(4) - 2 = 2(4) + 6$$

$$16 + 5 - 7 = 8 + 6$$

$$14 = 14$$

Because the statement $14 = 14$ is true, the answer $x = 4$ must be correct.

### 3.7.3 SETTING UP EQUATIONS

The equations discussed in the preceding paragraphs were expressed in *algebraic* language. You must learn how to set up an equation by translating a sentence into an equation (into algebraic language) and then solve the equation. The following suggestions and examples should be of help:

(1) Always read the statement of the problem carefully.
(2) Select the unknown number and represent it by some letter. If more than one unknown quantity exists in the problem, try to represent those numbers in terms of the same letter—that is, in terms of one quantity.
(3) Develop the equation, using the letter or letters selected, and then solve.

### *Example 6*

Five more than three times a number is the same as ten less than six times the number. What is the number?

*Solution:* Let $n$ represent the number.

$$3n + 5 = 6n - 10$$

$$3n - 6n = -10 - 5$$

$$-3n = -15$$

$$n = 5$$

**Example 7**

If five times the sum of a number and six is increased by three, the result is two less than ten times the number. Find the number.

*Solution*: Let $n$ represent the number.

$$5(n+6)+3 = 10n-2$$
$$5n+33 = 10n-2$$
$$5n = 10n-35$$
$$-5n = -35$$
$$n = 7$$

**Example 8**

The greater of two numbers is three less than seven times the smaller. Also, twelve more than the greater number is the same as ten times the smaller number. Find both numbers.

$$7n-3+12 = 10n$$
$$7n+9 = 10n$$
$$7n-10n = -9$$
$$-3n = -9$$
$$n = 3$$

The smaller number is 3.

$$7(n)-3$$
$$7(3)-3 = 18$$

The greater number is 18.

## 3.8  RATIO AND PROPORTION

### 3.8.1  RATIO

*Ratio* is the comparison of two numbers by division or an indicated division. The ratio of one number to another is determined when one number is divided by the other.

A ratio always includes two numbers. For example, if a box has a length of 8 in. and a width of 4 in., the ratio of the length to the width is expressed as 8/4 or 8:4. Both expressions have the same meaning.

All ratios are reduced to the lowest possible terms, similar to reducing a fraction to the lowest possible terms. The ratio of 8:4 should be reduced by dividing the 8 and the 4 by 4. The resulting ratio is 2:1. The ratio of the length of the box to its width is 2:1, since the box is 2 times as long as it is wide.

This ratio can also be stated as the relationship of the width to the length. The box is 4 in. wide and 8 in. long. The ratio of the width to the length is 4:8 or 4/8. This ratio, when reduced, becomes 1:2 or 1/2. The width of the box is 1/2 its length.

Let's look at an example of how and where ratios are used in air science work. In the U.S., concentrations of atmospheric gases and vapors are usually expressed as *mixing ratios* and reported in parts per million volume (ppmv). One ppmv is equal to one volume of a gas in 1 million volumes of air. (*Note:* Mixing ratios used to express air concentrations should not be confused with those used for water, which are weight/volume ratios (mg/L), and for solids, which are weight/weight (µg/gm, mg/km) ratios. Although all are expressed as ppm, they are not equivalent concentrations.)

$$1 \text{ ppmv} = \frac{1 \text{ gas volume}}{10^6 \text{ air volumes}}$$

A microliter volume of gas mixed in a liter of air would therefore be equal to 1 ppmv.

$$1 \text{ ppmv} = \frac{1 \,\mu\text{L gas}}{1 \text{ L air}}$$

Mixing ratios based on volume/volume ratios may also be expressed as parts per hundred million (pphmv), parts per billion (ppbv), or parts per trillion (pptv).

### 3.8.2 PROPORTIONS

Simply put, a *proportion* is a statement of equality between two ratios. Thus 2:4 = 4:8 is a proportion. We know that the two ratios are equal when the proportion is written in fractional form: 2/4 = 4/8. Either form may be read as follows: "Two is to four as four is to eight."

A general statement of the preceding proportion would be

$$\frac{a}{b} = \frac{c}{d} \text{ (fractional form)}$$

$$a:b = c:d \text{ (proportional form)}$$

where *a*, *b*, *c*, and *d* represent numbers.

A proportion may be written with a double colon (::) in place of the equal sign:

$$a:b::c:d$$

The first and last terms are called the extremes; the second and third terms are called the means.

When you consider proportions as fractions, it becomes evident that if any three members are known, then the fourth can be determined. When the proportion given earlier, $a:b = c:d$, is written as a fraction, $a/b = c/d$, then $a \times d = c \times b$. If you substitute numbers for letters in the proportion, then for $1/2 = 2/4$, $2 \times 2 = 4$, and $1 \times 4 = 4$. It is, therefore, obvious that the product of the means ($2 \times 2$) equals the product of the extremes ($1 \times 4$).

extremes                    extremes

$$a:b = c:d \qquad\qquad 1:2 = 2:4$$

means                        means

### Example 9

What is *x* in the proportion $2:3 = x:12$?

*Solution:* Rewriting in fractional form,

$$\frac{2}{3} = \frac{x}{12}$$

Since $12 = 3 \times 4$, *x* must be $2 \times 4$, or 8, because this gives equal fractions. Another method for solving this proportion is to use the principle that the product of the means equals the product of the extremes. The product of the means, $3x$, equals the product of the extremes, $2 \times 12$, or 24. Thus $3x = 24$, and *x* equals 8, the same answer obtained earlier.

### Example 10

What is *x* in the proportion $5:x = 2,000:10,000$?

*Solution:* Rewrite in fractional form,

$$\frac{5}{x} = \frac{2,000}{10,000}$$

and solve for the unknown value

$$5 = \frac{(2,000)(x)}{10,000}$$

$$(5)(10,000) = (2,000)(x)$$

$$\frac{(5)(10,000)}{2,000} = x$$

$$25 = x$$

## Example 11

If a pump will fill a tank in 20 hours at 4 gpm (gallons per minute), how long will it take a 10-gpm pump to fill the same tank?

First, analyze the problem. Here the unknown is some number of hours, but should the answer be larger or smaller than 20 hours? If a 4-gpm pump can fill the tank in 20 hours, a larger pump (10-gpm) should be able to complete the filling in less than 20 hours. Therefore, the answer should be less than the 20 hours.

Now set up the proportion:

$$\frac{x \text{ hours}}{20 \text{ hours}} = \frac{4 \text{ gpm}}{10 \text{ gpm}}$$

$$x = \frac{(4)(20)}{10}$$

$$x = 8 \text{ hours}$$

It doesn't take long until you gain an understanding of proportion problems that will allow you to skip some of the various steps in solving these problems (practice makes perfect and repetition aids easy recognition). In the following examples, a short-cut method is shown that will allow an experienced operator to solve problems quite easily.

## Example 12

To make a certain chemical solution, 66.3 mg of chemical must be added to 150 L of water. How much of the chemical should be added to 25 L to make up the same strength solution?

To solve this problem, you must first decide what is unknown and whether you expect the unknown value to be larger or smaller than the known value of the same unit. The amount of chemical to be added to 25 L is the unknown, and you would expect this to be smaller than the 66.3 mg needed for 150 L.

First, take the two known quantities of the same unit (25 L and 150 L) and make a fraction to multiply the third known quantity (66.3 mg) by. Notice that you can make two possible fractions with 25 and 150:

$$\frac{25}{150} \text{ or } \frac{150}{25}$$

Next, choose the fraction that will make the unknown number of milligrams less than the known. Multiplying by the fraction 25/150 would result in a number smaller than 66.3.

$$\frac{25}{150}(66.3) = z$$

$$\frac{(25)(66.3)}{150} = z$$

$$11.05 \text{ mg} = z$$

From the above operation it should be obvious that the key to this method is arranging the two known values of like units into a fraction that, when multiplied by the third known value, will render a result that is smaller or larger, as required.

### Example 13

If a machine metal is composed of 30 parts copper and 10 parts tin, what is the weight of each in a machine weighing 2,400 lb?

*Solution:* The total weight of the machine, in parts, is equal to 10 + 30, or 40 parts. The ratio of the number of parts of each metal to the total number of parts of metal equals the ratio of the weight of each metal to the total weight of metal. To calculate the pounds of copper, set up the following ratio:

$$\frac{30}{40} = \frac{c}{2,400} \text{ or } 30{:}40 = c{:}2,400$$

Then

$$30 \times 2,400 = c \times 40$$

$$c = \frac{30 \times 2{,}400}{40}$$

$$c = 1{,}800\,\text{lb}$$

For tin,

$$\frac{10}{40} = \frac{t}{2{,}400} \quad \text{or} \quad 10{:}40 = t{:}2{,}400$$

$$10 \times 2{,}400 = t \times 40$$

$$t = \frac{10 \times 2{,}400}{40}$$

$$t = 600\,\text{lb}$$

### 3.9  FINDING AVERAGES

Finding *averages* (or an arithmetic mean of a series of numbers) is accomplished by adding the numbers and dividing by the number of numbers in the group. This activity is required on several computations made by air specialists (e.g., field studies, statistical analysis, etc.).

Averaging plays an important role in scientific analysis. Averaging allows you to group data (information) and then compute an average from which a trend or trends may be determined. Note that an average is a reflection of the general nature of a certain group, and does not necessarily reflect any one component of that group.

***Example 14***

Find the average of the following series of numbers: 12, 14, 11, 6, 2, 9, 8, 7, 8, and 4. Adding the numbers together, we get 81. There are 10 numbers in this set so we divide 81 by 10 to get 8.1 as the average of the set.

### 3.10  PERCENT

Simply put, *percent* (%) means "parts of 100 parts" or "by the hundred." Percent is used to describe portions of the whole. Thus, 12% means 12 percent or 12/100 or 0.12. As another example, consider a tank that is 6/10 full; we say that it contains 60% of the original solution or of its total capacity. Percent is also commonly used to describe the portion of a budget spent on a project completed. "There is only 15% of the budgeted amount remaining." "The air sampling study is 40% complete."

Except when it is used in calculation, percentage is expressed as a whole number with a percent sign (%) after it. In a calculation, percent is expressed as a decimal. The decimal is obtained by dividing the percent by 100. For example, 12% is expressed as the decimal 0.12, since 12% is equal to 12/100. This decimal is obtained by dividing 12 by 100.

To determine what percentage a part is of the whole, divide the part by the whole. For example, if there are 110 sample blanks to label and Nancy has finished 40 of them, what percentage of the blanks have been labeled?

Step 1:  $40 \div 110 = 0.36$
Step 2:  The 0.36 is converted to percent by multiplying by 100.
Step 3:  $0.36 \times 100 = 36\%$. Thus, 36% of the 110 sample blanks have been labeled.

To determine the whole when the part and its percentage are given, divide the part by the percentage. Example: How much 65% calcium hypochlorite is required to obtain 15 pounds of chlorine? The part is 15 pounds, which is 65% of the whole.

Step 1:  Convert the percentage to a decimal by dividing by 100.
      $65\% \div 100 = 0.65$
Step 2:  Divide the part by the decimal equivalent of the percentage.
      15 pounds $\div 0.65 = 23.1$ (rounded)

To increase a value by a percent, we add the decimal equivalent of the percent to "1" and multiply it by the number.

A filter bed will expand 20% during backwash. If the filter bed is 48 inches deep, how deep will it be during backwash?

Step 1:  Change the percent to a decimal.
      $20\% \div 100 = 0.20$
Step 2:  Add the whole number 1 to this value.
      $1 + 0.20 = 1.20$
Step 3:  Multiply times the value.
      48 in. $\times 1.20 = 57.6$ inches

In air science work, the concentration of chemicals used or detected is commonly expressed as a percentage. For the sake of simplicity, let's look at an example using a liquid chemical solution.

Let's say we have a sulfur dioxide (an air pollutant in gaseous form) solution made to have a 6% concentration. It is often desirable to determine this concentration in mg/L. To accomplish this, we consider 6% as six percent of a million. [*Note*: A million is used because a liter of water weighs 1,000,000 mg. 1 mg in 1 L is 1 part in a million parts (ppm).] To find the concentration in mg/L when it is expressed in percent, do the following:

Step 1:  Change the percent to a decimal.
            6% ÷ 100 = 0.06
Step 2:  Multiply times a million.
            0.06 × 1,000,000 = 60,000 mg/L

### *Example 15*

Twenty percent of the chemical insecticide in a 650-gallon vat has been used. How many gallons are remaining in the vat?

Step 1:  Find the percentage of the insecticide remaining.
            100% − 20% = 80%
Step 2:  Change the percent to a decimal.
            80% ÷ 100 = 0.80
Step 3:  Multiply the percent as a decimal times the tank volume.
            0.80 × 650 gal = 520 gallons remain in the vat.

### 3.11  DEFINITIONS

Air sampling and air pollution control technologies may require you to perform calculations to determine the circumference, area, or volume of ventilation systems ducting, tanks, vessels, and other structures. You may also need to calculate the perimeter of landscapes as part of field data necessary to determine the result or the correct parameters to set or to measure in any gaseous or volatile chemical spill mitigation action. To aid in performing these calculations, the following definitions are provided:

- *Area*   the amount of surface within the lines of a geometric figure, measured in square units.
- *Base*   the bottom leg of a triangle, measured in linear units.
- *Circumference*   the distance around an object, measured in linear units. When determined for other than circles, it may be called the *perimeter* of the figure, object, or landscape.
- *Cubic units*   measurements used to express volume—cubic feet, cubic meters, etc.
- *Depth*   the vertical distance from the bottom of the tank to the top. Normally measured in terms of liquid depth and given in terms of side wall depth (SWD), measured in linear units.
- *Diameter*   the distance from one edge of a circle to the opposite edge, passing through the center; measured in linear units.

- *Height*   the vertical distance from the base or bottom of a unit to the top or surface.
- *Length*   the distance from one end of an object to the other, measured in linear units.
- *Linear units*   measurements used to express distances—feet, inches, meters, yards, etc.
- *Pi, π*   a number in the calculations involving circles, spheres, or cones. π = 3.1416.
- *Radius*   the distance from the center of a circle to the edge, measured in linear units.
- *Sphere*   a container shaped like a ball.
- *Square units*   measurements used to express area—square feet, square meters, acres, etc.
- *Volume*   the capacity of an object (how much it will hold), measured in cubic units (cubic feet, cubic meters) or in liquid volume units (gallons, liters, million gallons).
- *Width*   the distance from one side of the object to the other, measured in linear units.

## 3.12  PERIMETER AND CIRCUMFERENCE

Air science practitioners may be called upon to make certain measurements in the field related to air pollution control technology or to make design considerations and other related activities. On occasion, they may need to determine the distance around grounds or landscapes. To measure the distance around property, buildings, and basin-like structures, you must first determine either perimeter or circumference. The *perimeter* is how far it is around an object or area such as a piece of ground. *Circumference* is the distance around a circle or circular object. Distance is a linear measurement, which defines the length along a line. Standard units of measurement (inches, feet, yards, and miles) and metric units (centimeters, meters, and kilometers) are used.

The circumference (*C*) of a circle is found by multiplying pi (π) times the diameter (*D*) (diameter is a straight line across the circle passing through the center).

$$C = \pi D \qquad\qquad (3.8)$$

where

$C$ = circumference
$\pi$ = Greek letter pi
$\pi$ = 3.14
$D$ = diameter

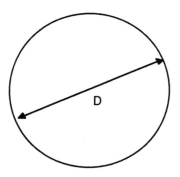

### Example 16

A circular chemical holding tank has a diameter of 16 m. What is the circumference of this tank?

$$C = \pi(D)$$
$$C = (3.14)\,(\text{diameter})$$
$$C = (3.14)\,(16\ \text{m})$$
$$C = 50.2\ \text{m}$$

### Example 17

A ventilation air test inlet opening has a diameter of 3 inches. What is the circumference of the inlet opening in inches?

$$C = \pi(D)$$
$$C = 3.14 \times 3\ \text{in.}$$
$$C = 9.42\ \text{in.}$$

## 3.13  AREA

For area measurements in air science work, three basic shapes are particularly important, namely circles, rectangles, and triangles.

*Area* is the amount of surface an object contains, or the amount of material it takes to cover the surface. The area on top of a chemical tank is called the *surface area*. The area of the end of a ventilation duct is called the *cross-sectional area* (the area at right angles to the length of ducting). Area is usually expressed in square units such as square inches (in.²) or square feet (ft²). Land may also be expressed in terms of square miles (sections) or acres (43,560 sq mi), or in the metric system as hectares.

The area of a rectangle is found by multiplying the length ($L$) times width ($W$).

$$\text{Area} = L \times W \tag{3.9}$$

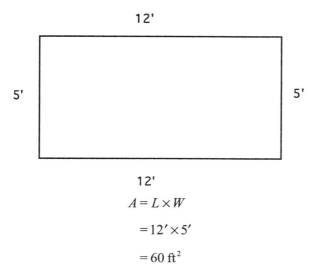

Find the area of the following rectangle:

$$A = L \times W$$

$$= 12' \times 5'$$

$$= 60 \text{ ft}^2$$

The surface area of a circle is determined by multiplying $\pi$ times the radius squared. Radius, designated $r$, is defined as a line from the center of a circle or sphere to the circumference of the circle or surface of the sphere.

$$\text{area of circle} = \pi r^2 \tag{3.10}$$

where:

$A$ = area
$\pi$ = Greek letter pi ($\pi = 3.14$)
$r$ = radius or a circle (radius is one-half the diameter)

### Example 18

What is the area of the circle shown below?

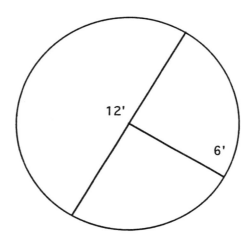

$$\text{Area of circle} = \pi r^2$$

$$= \pi 6^2$$

$$= 3.14 \times 36$$

$$= 113 \text{ ft}^2$$

## 3.14 VOLUME

The amount of space occupied by or contained in an object, *volume*, is expressed in cubic units, such as cubic inches (in.³), cubic feet (ft³), acre feet (1 acre foot = 43,560 ft³), etc.

The volume of a rectangular object is obtained by multiplying the length times the width times the depth or height.

$$V = L \times W \times H \tag{3.11}$$

where

$L = $ length
$W = $ width
$D$ or $H = $ depth or height

For example, find the volume in cubic feet of a holding pond with the following dimensions: length, 15 ft; width, 7 ft; and depth, 9 ft.

$$V = L \times W \times D$$
$$= 15' \times 7' \times 9'$$
$$= 945 \text{ ft}^3$$

For air science practitioners, representative surface areas are most often rectangles, triangles, circles, or a combination of these. Practical volume formulas used in air science calculations are given in Table 3.3.

***Example 19***

Find the volume of a 4-inch round air duct that is 300 feet long.

Step 1: Change the diameter of the duct from inches to feet by dividing by 12.

$$D = 4 \div 12 = 0.33 \text{ ft}$$

Step 2: Find the radius by dividing the diameter by 2.

$$r = 0.33 \text{ ft} \div 2 = 0.165$$

TABLE 3.3. Volume Formulas.

| | |
|---|---|
| Sphere volume | = $(\pi/6)$ (diameter)$^3$ |
| Cone volume | = 1/3 (volume of a cylinder) |
| Rectangular tank volume | = (area of rectangle) (*D* or *H*) |
| | = (*LW*) (*D* or *H*) |
| Cylinder volume | = (area of cylinder) (*D* or *H*) |
| | = $\pi r^2$ (*D* or *H*) |
| Triangle volume | = (area of triangle) (*D* or *H*) |
| | = (*bh*/2) (*D* or *H*) |

Step 3: Find the volume

$$V = L \times \pi r^2$$
$$V = 300 \text{ ft} \times 3.14 \times (.0272) \text{ ft}^2$$
$$V = 25.6 \text{ ft}^3$$

## Example 20

Find the volume of a smoke stack that is 36 inches in diameter (its entire length) and 96 inches tall.

Step 1:  Find the radius of the stack, which is one-half the diameter.

$$36 \text{ inches} \div 2 = 18 \text{ inches}$$

Step 2:  Find the volume

$$V = H \times \pi r^2$$
$$V = 96 \text{ in.} \times \pi (18)^2$$
$$V = 96 \text{ in.} \times \pi (324 \text{ in.}^2)$$
$$V = 97,667 \text{ in.}^3$$

## 3.15  CONVERSION FACTORS

Conversion factors are used to change measurements or calculated values from one unit of measure to another. In making the conversion from one unit to another, you must know two things:

(1)  The exact number that relates the two units
(2)  Whether to multiply or divide by that number

For example, in converting from inches to feet, you must know that there are 12 in. in 1 ft, and you must know whether to multiply or divide the number of inches by 0.08333 (i.e., 1 in. = 0.08 ft).

When people make conversions, they often become confused about whether to multiply or divide, although the number that relates the two units is usually known and thus is not a problem. Gaining an understanding of the proper methodology—the "mechanics"—to use for various operations requires practice.

Along with using the proper "mechanics" (and much practice) in making conversions, probably the easiest and fastest method of converting units is to use a conversion table. Many experienced air practitioners memorize some standard conversions. This occurs as a result of on-the-job perform-ance, practice, and repetition, not normally as a result of attempting to memorize them. There is a definite need for conversion tables.

As an example of the type of conversions the air science practitioner must be familiar with, we provide the following section on temperature conver-sions.

### 3.15.1 TEMPERATURE CONVERSIONS

A conversion is a number used to multiply or divide into another number in order to change the units of the number. In many instances, the conversion factor cannot be derived—it must be known.

Most air science practitioners are familiar with the formulas used for Fahrenheit and Celsius temperature conversions:

$$^{\circ}C = 5/9 \, (^{\circ}F - 32)$$

$$^{\circ}F = 9/5 \, (^{\circ}C) + 32$$

The difficulty arises when one is required to recall these formulas from memory. Probably the easiest way to make temperature conversions is to remember these three basic steps for both Fahrenheit and Celsius conver-sions:

(1) Add 40°.
(2) Multiply by the appropriate fraction (5/9 or 9/5).
(3) Subtract 40°.

Obviously, the only variable in this method is the choice of 5/9 or 9/5 in the multiplication step. To make the proper choice, you must be familiar with the two scales. As shown in Figure 3.2, on the Fahrenheit scale the freezing point of water is 32° whereas it is 0° on the Celsius scale. The boiling point of water is 212° on the Fahrenheit scale and 100° on the Celsius scale.

Note that at the same temperature, higher numbers are associated with the Fahrenheit scale and lower numbers with the Celsius scale. This is an important relationship that helps you decide whether to multiply by 5/9 or 9/5.

### *Example 21*

Suppose that you wish to convert 220°F to Celsius. Using the three-step process we proceed as follows:

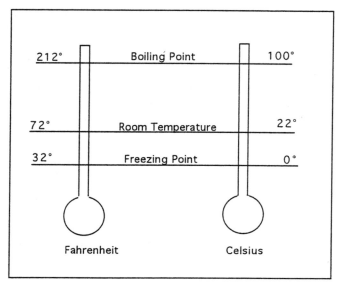

**Figure 3.2** Fahrenheit and Celsius temperature scales.

Step 1:  Add 40°.

$$220° + 40° = 260$$

Step 2:  260° must be multiplied by either 5/9 or 9/5. Since the conversion is to the Celsius scale, you will be moving to a smaller number. Through reason and observation it is obvious that, if 260 is multiplied by 9/5, it would almost double 240 rather than make it smaller. On the other hand, if you multiply by 5/9, it would almost cut 260 in half. Since in this problem you wish to move to a smaller number, you should multiply by 5/9:

$$(5/9)(260°) = 144.4°C$$

Step 3:  Now subtract 40°.

$$144.4°C - 40.0°C = 104.4°C$$

Therefore, 220°F = 104.4°C

When converting from Celsius to Fahrenheit, you are moving from a smaller to a larger number, so 9/5 should be used in the multiplication.

Obviously, knowing how to make these temperature conversion calculations is useful. However, in practical operations you may wish to use a temperature conversion table.

## 3.16 SUMMARY

Remember that we have only touched on the type of mathematics that you will be called upon to use in air science in this brief review of the fundamentals. Calculus is the common tool of the engineer and others who design air pollution control technology and other air-related systems. However, this text is designed to present the big picture in a clear, unfettered way—without the technicalities of higher math.

In the following chapters many theoretical and practical situations will be illustrated where math plays an important role.

## 3.17 REFERENCES

Heinlein, R. A., *Time Enough for Love*. New York: G.P. Putnam, 1973.
Price, J., *Basic Math Concepts for Water and Wastewater Plant Operators*. Lancaster, PA: Technomic Publishing Co., Inc., 1991.

# CHAPTER 4

# Air Physics

*The difference between science and the fuzzy subjects is that science requires reasoning, while those other subjects merely require scholarship. (Heinlein, 1973, p. 348)*

## 4.1 INTRODUCTION

$I$T was pointed out earlier that the air science practitioner must be well grounded in science fundamentals. Although the environmental engineer usually designs the air pollution control equipment, if you are responsible for the proper operation and selection of such equipment, it is incumbent upon you to know as much as possible about air, the ingredients that make up air, and the flow of air.

A knowledge of air physics (or more correctly stated, the physics of gases and particles) is fundamental to gaining an understanding of the characteristics of a system in which control equipment is used. For example, if you are required to select air-handling equipment (including fans, simple duct work, and collection equipment), you must be able to determine the volume of air to be handled.

In addition, you will need to have a good working knowledge of a few basic physical properties such as gas density, pressure drop, viscosity of the gas, and the pressure drop in filter media. This chapter discusses the basic physical properties critical to gaining a better understanding of equipment operation. Because their application is important to air quality and emissions sampling/monitoring, emission assessment procedures, data summarization, engineering controls, and air quality monitoring, special attention is given to the gas laws.

## 4.2 FORCE, WEIGHT, AND MASS

In air science work, the properties of gases must be known for proper selection and operation of air pollution control devices and ancillary equipment (ductwork). Along with knowing the properties of gases, the air science practitioner must also be familiar with the vocabulary and understand the meaning of (as well as the difference between) the terms force, weight, and mass.

*Force* is any influence that tends to change a body's state of rest or its uniform motion in a straight line. Stated in simpler terms, force is a push or a pull exerted on an object to change its position or movement, including starting, stopping, and changing its speed or direction of movement. In a compressed air (pneumatic) system, force must be present at all times for the system to function.

A substance or object has *weight* depending on its *mass* (represents the amount of matter in an object and its *inertia,* or resistance to movement) and the strength of the Earth's gravitational pull, which decreases with height. Consequently, an object weighs less at the top of a mountain than at sea level. An object's inertia determines how much force is needed to lift or move the object or to change its speed or direction of movement.

Another important physical property is *density,* which is a scalar quantity. The density of an object is its weight for a specific volume or unit of measure, a measure of the compactness of a substance. Density is equal to its mass per unit volume and is measured in kilograms per cubic meter or pounds per cubic foot. The density of a mass $m$ occupying a volume $V$ is given by the formula

$$D = m/V \qquad (4.1)$$

The density of a cubic foot of dry air at atmospheric pressure and a temperature of 60°F is 0.076 lb and is more commonly expressed as 0.076 lb/ft$^3$. The density of wet air at atmospheric pressure with 100% relative humidity and a temperature of 60°F is 0.075 lb/ft$^3$. Humid air is less dense than dry air because the water vapor will not allow the air to compress as much. As a result, humid air weighs less. Air's relatively low density makes

TABLE 4.1. Densities of Gases at STP (standard temperature and pressure: 0°C and 1 atm).

| | |
|---|---|
| Air | 1.3 |
| Hydrogen | 0.09 |
| Helium | 0.18 |
| Methane | 0.72 |
| Nitrogen | 1.25 |
| Oxygen | 1.43 |
| Carbon dioxide | 1.98 |
| Propane | 2.02 |
| Butane | 2.65 |

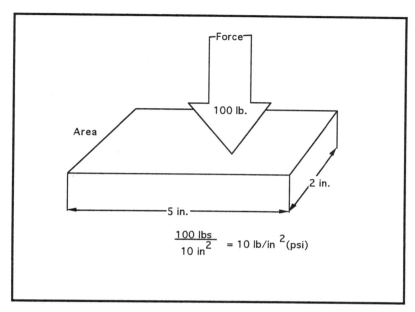

**Figure 4.1** Determining pressure.

it suitable for long-distance and high-speed control applications in pneumatic systems.

The densities of some common gases are given in Table 4.1.

### 4.3 PRESSURE

*Pressure* is the amount of force (in pounds) exerted on an object or a substance, divided by the area (in square inches) over which this force is exerted. Pressure can be measured and specified in different ways but is commonly measured in pounds per square inch (psi). The SI unit of pressure is the pascal (newton per square meter), equal to 0.01 millibars. At the outer edge of Earth's atmosphere, pressure is zero, whereas at sea level, *atmospheric pressure* (because of the weight of the air above) is about 100 kilopascals (1,013 millibars or 1 atmosphere).

Let's look at a means of determining pressure.

Figure 4.1 shows a 100-lb force applied to an area of 10 in.$^2$; the resulting pressure is 10 lb/in.$^2$ (psi). If the pressure (in psi) on a certain area (square inches, feet, etc.) is known, the total force (in pounds) exerted by the pressure is equal to the pressure multiplied by the area.

In this book we are primarily concerned with atmospheric pressure, but note that two other kinds of pressure—below atmospheric and pneumatic system pressure—are also common. Atmospheric pressure at sea level equals 14.7 psi; pressure is lower above sea level and higher below sea level.

The complete or partial absence of air (indicating below atmospheric pressure) is often referred to as a *vacuum* or partial vacuum. In some applications, it may also be called a negative or suction pressure. Vacuum is normally measured using special gages or with a column of mercury. When all the air above the column is evacuated, atmospheric pressure is exerted on the pool of mercury below the tube. This pressure raises the column to a height of approximately 30 in. In most applications a vacuum is measured in inches of mercury instead of psi.

Note that most pressure gages in a pressurized air system measure only pressure that is higher than the atmospheric pressure surrounding them. You may have noticed that, when a pressure gage is disconnected, it reads zero pounds per square inch, which is known as *gage pressure* (0 psig). For example, a reading of 300 on an air system pressure gage tells you that the air pressure is 300 psi above atmospheric. If we add atmospheric pressure to this gage pressure, the total pressure is 314.7 pounds per square inch (300 + 14.7), which is known as *absolute pressure* (psia). Remember that although absolute pressure readings are important in some pressurized air system calculations, the distinction between psig and psia is usually unimportant in the average air system. As a result, gage pressure readings are usually expressed in psi.

### 4.4 WORK AND ENERGY

*Work* is the transference of energy that occurs when a force is applied to a body that is moving in such a way that the force has a component in the direction of the body's motion. Stated in simpler fashion: Work takes place when a force (in pounds or newtons) moves through a distance (in inches, feet, or meters). The amount of work done is expressed in the English system of measurement in foot-pounds or inch-pounds, as shown in the following equation:

$$\text{work} = \text{force (pounds)} \times \text{distance (feet or inches)} \qquad (4.2)$$

$$= \text{foot-pounds or inch-pounds}$$

In a pressurized air (pneumatic) system, the force in pounds is exerted by air pressure acting on the area of a moving piston in a cylinder (see Figure 4.2). As the piston moves, the pneumatic force acts through the length of the stroke. You can determine the work done by the piston by using the following equations:

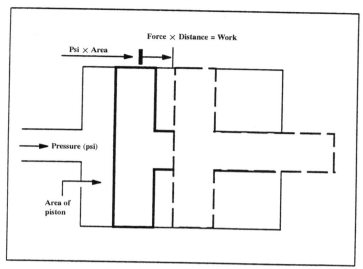

**Figure 4.2** Pressurized air at work.

$$\text{force }(F) = \text{pressure} \times \text{area (piston area)} \qquad (4.3)$$
$$= \frac{\text{lb}}{\text{in}^2} \times \text{in}^2$$
$$= \text{lb}$$

$$\text{work }(W) = \text{force} \times \text{distance (piston travel)} \qquad (4.4)$$
$$= \text{lb} \times \text{in.}$$
$$= \text{in.-lb}$$

To convert to foot-pounds, divide in.-lb by 12:

$$\frac{\text{in.-lb}}{12} = \text{foot-pounds}$$

*Power* is defined as the time rate of doing work or as the amount of work (foot-pounds) done in a given length of time (seconds or minutes) in foot-pounds per minute. The following equation is used to determine the amount of power:

$$\text{power }(P) = \frac{\text{work}}{\text{time}} \qquad (4.5)$$
$$= \frac{\text{foot-pounds}}{\text{seconds (or minutes)}}$$

To convert foot-pounds to inch-pounds, multiply by 12.

Note that, for the amount of power calculated to be meaningful, it must be compared with a unit of measurement. The common unit of power measurement is *horsepower*, calculated as follows:

$$1 \, hp = \frac{33,000 \, ft\text{-}lb}{seconds \, (or \, minutes)} \tag{4.6}$$

When power is used to perform work, *energy* is expended. The *Law of Conservation of Energy* states, "Energy cannot be created or destroyed. It can only be transformed." Thus, we use one kind of energy to get other kinds of energy. Some of this energy does useful work while some of it is wasted as heat energy in overcoming friction.

## 4.5 DIFFUSION AND DISPERSION

*Diffusion* is the spontaneous and random movement of molecules or particles in a gas (or liquid) from a region in which they are at a high concentration until a uniform concentration is achieved throughout. No mechanical mixing or stirring is involved. This should not be confused with evaporation, which is the changing of a liquid to a gas.

*Dispersion* is the temporary mixing of liquid particles with a gas.

Diffusion and dispersion are important in air pollution. For example, in dispersion, air pollutants are diluted and reduced in concentration. Air pollution dispersion mechanisms are a function of the prevailing meteorological conditions. Diffusion and dispersion (in air pollution) will be discussed much more fully in Chapter 15.

## 4.6 COMPRESSIBILITY

Air, unlike liquids, is readily compressible, and large quantities can be stored in relatively small containers. The more the air is compressed, the higher its pressure becomes. The higher the pressure in a container, the stronger the container must be. Gases are important compressible fluids, not only from the standpoint that a gas can be a pollutant, but also because gases convey the particles (particulate matter) and gaseous pollutants (Hesketh, 1991).

## 4.7 GAS LAWS

Gases can be pollutants as well as the conveyors of pollutants. Air (which is mainly nitrogen) is usually the main gas stream. Gas conditions are usually described in two ways: *standard temperature and pressure (STP)* and *Standard Conditions (SC)*. STP represents 0°C (32°F) and 1 atm. SC is more commonly used and represents typical room conditions of 20°C (70°F) and 1 atm; SC is usually measured in cubic meters, Nm³, or standard cubic feet (scf).

To understand the physics of air, it is imperative that you have an understanding of various physical laws that govern the behavior of pressurized gases. One of the more well-known physical laws is *Pascal's Law,* which states that a confined gas (fluid) transmits externally applied pressure uniformly in all directions without change in magnitude (see Figure 4.3). If the container is flexible, it will assume a spherical (balloon) shape (Figure 4.3). However, most compressed-gas tanks are cylindrical in shape (which allows use of thinner sheets of steel without sacrificing safety) with spherical ends to contain the pressure more effectively.

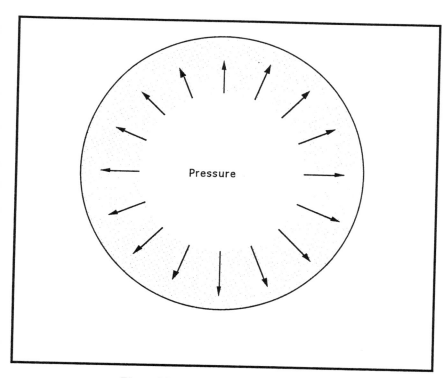

**Figure 4.3** Action of confined air pressure.

## 4.7.1 BOYLE'S LAW

Though gases are compressible, note that, for a given mass flow rate, the actual volume of gas passing through the system is not constant within the system because of changes in pressure. This physical property (the basic relationship between the pressure of a gas and its volume) is described by *Boyle's Law* (named for its discoverer, Irish physicist and chemist Robert Boyle, in 1662), which states: "The absolute pressure of a confined quantity of gas varies inversely with its volume at a given temperature." For example, if the pressure of a gas doubles, its volume will be reduced by half, and vice versa. This means, for example, that if 12 ft³ of air at 14.7 psia is compressed to 1 ft³, air pressure will rise to 176.4 psia, as long as air temperature remains the same. Figure 4.4 shows this relationship, which can be calculated as follows:

$$P_1 \times V_1 = P_2 \times V_2 \tag{4.7}$$

where

$P_1$ = original pressure (units for pressure must be absolute)
$V_1$ = new pressure (units for pressure must be absolute)
$P_2$ = original gas volume at pressure $P_1$
$V_2$ = new gas volume at pressure $P_2$

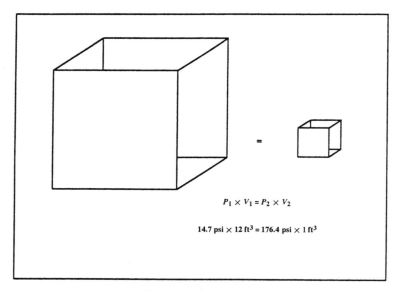

$$P_1 \times V_1 = P_2 \times V_2$$

$$14.7 \text{ psi} \times 12 \text{ ft}^3 = 176.4 \text{ psi} \times 1 \text{ ft}^3$$

**Figure 4.4** Boyle's Law.

This equation can also be written as:

$$\frac{P_2}{P_1} = \frac{V_1}{V_2} \text{ or } \frac{P_1}{P_2} = \frac{V_2}{V_1} \tag{4.8}$$

To allow for the effects of atmospheric pressure, always remember to convert from gage pressure *before* solving the problem, and then convert back to gage pressure *after* solving it:

$$\text{psia} = \text{psig} + 14.7 \text{ psi}$$

and

$$\text{psig} = \text{psia} - 14.7 \text{ psi}$$

Note that in a pressurized gas system (where gas is caused to move through the system by the fact that gases will flow from an area of high pressure to that of low pressure), we will always have a greater actual volume of gas at the end of the system than at the beginning (assuming the temperature remains constant).

Let's take a look at a typical gas problem using Boyle's Law.

### Example 22

What is the gage pressure of 12 ft³ of air at 25 psig when compressed to 8 ft³?

*Solution:*

$$25 \text{ psig} + 14.7 \text{ psi} = 42.7 \text{ psia}$$

$$P_2 = P_1 \times \frac{V_1}{V_2} = 42.7 \times \frac{12}{8} = 64 \text{ psia}$$

$$\text{psig} = \text{psia} - 14.7 \text{ psi}$$

$$= (64 \text{ psia}) - (14.7 \text{ psi}) = 49.3 \text{ psig}$$

The gage pressure is 49.3 psig (remember that the pressures should always be calculated on the basis of absolute pressures instead of gage pressures).

### 4.7.2 CHARLES'S LAW

Another physical law dealing with temperature is *Charles's Law* (discovered by French physicist Jacques Charles in 1787). It states, "the volume of a given mass of gas at constant pressure is directly proportional to its absolute temperature." [The temperature should be in Kelvin (273 + °C) or Rankine (absolute zero = −460°F or 0°R).]

This is calculated by using the following equation:

$$P_2 = P_1 \times \frac{T_2}{T_1} \tag{4.9}$$

Charles's Law also states: "If the pressure of a confined quantity of gas remains the same, the change in the volume ($V$) of the gas varies directly with a change in the temperature of the gas," as given in the equation:

$$V_2 = V_1 \times \frac{T_2}{T_1} \tag{4.10}$$

### 4.7.3 IDEAL GAS LAW

The *Ideal Gas Law* combines Boyle's and Charles's Laws because air cannot be compressed without its temperature changing. The Ideal Gas Law is expressed by the equation:

$$\frac{P_1 \times V_1}{T_1} = \frac{P_2 \times V_2}{T_2} \tag{4.11}$$

Note that the Ideal Gas Law is still used as a design equation even though the equation shows that the pressure, volume, and temperature of the second state of a gas are equal to the pressure, volume, and temperature of the first state. In actual practice, however, other factors (humidity, heat of friction, and efficiency losses, for example) affect the gas. This equation uses absolute pressure (psia) and absolute temperatures (°R) in its calculations.

## 4.8 HEAT AND ENERGY IN THE ATMOSPHERE

The sun's energy is the prime source of Earth's climatic system. From the sun, energy is reflected, scattered, absorbed, and reradiated within the system but without uniform distribution. Some areas receive more energy than they lose; in other areas, the reverse occurs. If this situation could continue for long, the areas with an energy surplus would get hotter—too hot—and those with a deficit would get colder—too cold. This does not happen because the temperature differences produced help to drive the world's wind and ocean currents. They carry heat with them (either in sensible or latent forms) and help to counteract the radiation imbalance. Winds from the tropics (normally warm) carry excess heat with them. Polar winds blow from areas with a heat deficit and so carry cold air. Acting together, these energy transfer mechanisms help to produce Earth's present climates.

## 4.9 ADIABATIC LAPSE RATE

The atmosphere is restless, always in motion either horizontally, vertically or both. As air rises, pressure on it decreases, and in response it expands. The act of expansion to encompass its new and larger dimensions requires an expenditure of energy; since temperature is a measure of internal energy, this use of energy makes its temperature drop. This is an important point and an important process in air physics.

This phenomenon is known as the *adiabatic lapse rate*. Simply stated, adiabatic refers to a process that occurs with or without loss of heat, especially the expansion or contraction of a gas in which a change takes place in pressure or volume, although no heat is allowed to enter or leave. Lapse rate refers to the rate at which air temperature decreases with height. The normal lapse rate in stationary air is on the order of 3.5°F/1000 ft (6.5°C/km). This value may vary with latitude and changing atmospheric conditions. A parcel of air that is not immediately next to the Earth's surface is sufficiently well insulated by its surroundings that either expansion or compression of the parcel may be assumed to be adiabatic.

The air temperature may be calculated for any height by the general formula

$$T = T_0 - Rh \qquad (4.12)$$

where

$T$ = temperature of the air
$h$ = height of air
$T_0$ = temperature of the air at the level from which the height is measured
$R$ = lapse rate

## *Example 23*

If the air temperature of stationary air ($R = 3.5°F/1000$ ft) at the Earth's surface is 70°F, then at 5000 ft, the stationary air temperature would be

$$T = T_0 - Rh$$

$$= 70°\,\text{F} - (3.5°\,\text{F}/1000\,\text{ft})\,(5000\,\text{ft})$$

$$= 70°\,\text{F} - 17.5°\,\text{F} = 52.5°\,\text{F}$$

The formula simply says that, for every 1000 ft of altitude (height), 3.5°F is subtracted from the initial air temperature in this case.

Adiabatic lapse rates have an important relationship with atmospheric stability and will be discussed in greater detail later.

## 4.10  THE ALBINO RAINBOW

Have you ever looked up into the sky and seen 11 suns? Have you been at sea and witnessed the towering, spectacular Fata Morgana? How about a "glory"—have you ever seen one? Or how about the albino rainbow? Do you know what these are? They will be described shortly.

Normally, when we look up into the sky, we see what we expect to see: an ever-changing backdrop of color with dynamic vistas of blue sky, white, puffy clouds, gray storms, and gold and red sunsets. On some occasions, however, when atmospheric conditions are just right, we can look up at the sky or out to the horizon and see strange phenomena (lights in the sky). What causes these momentary wonders?

Because Earth's atmosphere is composed of gases (air), it is actually a sea of molecules. These molecules of air scatter the blue, indigo, and violet shorter wavelengths of light more than the longer orange and red wavelengths, which is why the sky appears blue.

What are wavelengths of light? Figure 4.5 will help answer this question. Simply put, wavelength of light actually refers to the electromagnetic spectrum. The portion of the spectrum visible to the human eye falls between the infrared and ultraviolet wavelengths.

The word *light* is commonly given to visible electromagnetic radiation. However, only the frequency (or wavelength) distinguishes visible electromagnetic radiation from the other portions of the spectrum.

Let's get back to looking up into the sky. Have you ever noticed right after a rain shower how dark a shade of blue the sky appears? Have you looked out upon the horizon at night or in the morning and noticed that the sun's light produces a red sky? This phenomena is caused by sunlight passing through large dust particles, which scatter the longer wavelengths. Have you ever noticed that fog and cloud droplets, with diameters larger than the wavelength of light, scatter all colors equally and make the sky look white? Maybe you have noticed that fleeting greenish light that appears just as the sun sets? It occurs because different wavelengths of light are *refracted* (bent)

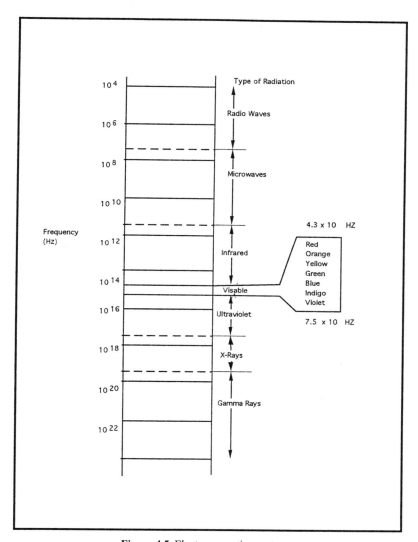

**Figure 4.5** Electromagnetic spectrum.

in the atmosphere by differing amounts. Because green light is refracted more than red light by the atmosphere, green is the last to disappear.

What causes rainbows? A rainbow is really nothing more than an airborne prism. When sunlight enters a raindrop, refraction and reflection take place, splitting white light into the spectrum of colors from red to blue and making a rainbow.

Earlier a "glory" was mentioned. Interactions of light waves can produce a glory. For example, if you were standing on a mountain, with the sun to your back, you may cast a shadow on the fog in the valley. Your shadow may appear to be surrounded by colored halos—a glory. The glory is caused by light entering the edges of tiny droplets and being returned in the same direction from which it arrived. These light waves interfere with each other, sometimes canceling out and sometimes adding to each other.

Why do we sometimes see multiple suns? Reflection and refraction of light by ice crystals can create bright halos in the form of arcs, rings, spots, and pillars. Mock suns (sun dogs) may appear as bright spots 22° to the left or right of the sun. Sun pillars occur when ice crystals act as mirrors, creating a bright column of light extending above the sun. Such a pillar of bright light may be visible even when the sun has set.

What is the Fata Morgana? It is an illusion that often fools sailors into seeing mountain ranges floating over the surface of the ocean. An albino rainbow is an eerie phenomenon that can only be seen on rare occasions in foggy conditions.

Atmospheric phenomena (lights in the sky) are real, apparent, and sometimes visible. Awe inspiring as they are, their significance and their actual existence are based on physical conditions that occur in our atmosphere.

## 4.11 PARTICLE PHYSICS

In air pollution control technology, you must have a basic understanding of the physics of gases and also of the physics of airborne particles or particulate matter (or aerosol) that might be entrained within the gaseous stream. Particulates constitute a major class of air pollution. They have a variety of shapes; they include dusts, fumes, mists, and smoke, and they have a wide range of physical and chemical properties.

Important particulate matter characteristics include size, size distribution, shape, density, stickiness, corrosivity, reactivity, and toxicity. With this wide range of characteristics, obviously, one type of collection device might be better suited than others for a particular particle. For example, particulate matter typically ranges in size from 0.005 to 100 microns in diameter. This wide range in size distribution must certainly be taken into account.

Particle size is based on particle behavior in the Earth's gravitational field. The aerodynamic equivalent diameter refers to a spherical particle of unit density that falls at standard velocity.

Particle size determines atmospheric lifetime, effects on light scattering, and deposition in human lungs. However, when we speak of particulate size, we are not referring to particulate shape. This seems obvious, but for simplicity and for theoretical and learning applications, common practice bases assumptions on particles being spherical. Particles are not often spherical; they are usually quite irregularly shaped. In pollution control technology design applications, the size of the particle is taken into consideration, but the particle's behavior in a gaseous waste stream is our chief concern. To better understand this important point, consider an analogy. If you take a flat piece of regular typing paper and drop it from the top of a six-foot ladder, the paper will slip and slide sideways on the air, settling irregularly to the floor. Crumple the same sheet of paper and drop it from the ladder, and the crumpled paper will fall rapidly. Yes, size is important, but other factors—including particle shape—are also important.

Along with particle size and shape, particle density must also be taken into consideration. For example, a baseball and a whiffle ball have approximately the same diameter but behave quite differently when you toss them into the air.

## 4.12 SUMMARY

Physical laws govern our lives, our world, and our universe. These laws are ancient, immutable, unyielding, and unchanging. We cannot alter them. We can only adapt ourselves to them and learn to use them to our advantage. The physical laws that govern air are critical to environmental science and engineering, and the basics of physics and chemistry, as they relate to air science, are critical to a solid environmental science education.

## 4.13 REFERENCES

Heinlein, R. A., *Time Enough for Love*. New York: G.P. Putnam, 1973.

Hesketh, H. E., *Air Pollution Control: Traditional and Hazardous Pollutants*. Lancaster, PA: Technomic Publishing Co,. Inc., 1991.

# CHAPTER 5

# Air Chemistry

*We are normally too engrossed with the activities of our daily lives to consider what life would be like without chemistry. If we were to ponder this point, however, it would soon be apparent that chemistry affects everything that we do. Not a single moment of time goes by which we are not affected somehow by a chemical substance or chemical process. (Eugene Meyer, 1989, p. 5)*

## 5.1 INTRODUCTION

CHEMISTRY—chemists—chemicals? As Meyer points out above, we normally give little thought to these terms. When we do, what do we think? Some folks immediately tense up when the term *chemistry* is mentioned. To them, chemistry is a strange term with stranger connotations.

There are many views of chemistry and chemists (in particular)—too many different perceptions to describe here. One thing is certain, however. When we mention chemistry to almost anyone, the image that usually forms is one of the mixing of "liquid" chemicals. People have the perception that chemistry is most often something that happens in liquids, especially the exotic liquids—the colored ones featured in horror stories, science fiction tales, and science classes. Is it any surprise that the average person would never equate chemistry to something occurring in the atmosphere as well? They do not know enough about chemistry to understand the multitude of reactions—slow and fast chemical reactions, dissolving chemicals, precipitation of colored solids—that occur in air.

Countless experiments have been conducted in Earth's atmosphere over the past few centuries. Consider, for example, the weather balloons that we have sent up with special measuring instruments. The instruments measure certain atmospheric conditions to aid us in forecasting the weather. These balloons have done much more for us than that. They have revealed that the atmosphere contains several thousand different chemical species.

Experiments have also been conducted by satellites circling the globe, and some airplanes have been outfitted as virtual flying chemical laboratories.

Experiments have been conducted on the Earth's surface as well, all of which have also helped to catalog thousands of different chemical species. Have we cataloged them all? No, not really. In chemistry and in science in general, we are always making new discoveries—always trying to find out something about the "we do not know what we do not know" syndrome.

But we do know a lot. For example, as Graedel and Crutzen (1995) put it, we know that Earth's atmosphere is literally a "flask without walls," containing several thousand different chemical species (p. 35). Many consist of the atmospheric gas itself; some are found in airborne particulate matter, both large and small, and some are found dissolved in hydrometeors. Many atmospheric compounds have both natural and anthropogenic origins. The compounds from natural origins (volcanic action, earthquakes, horrendous storms, and others) have been with us from the very beginning. Anthropogenic contributions are more recent—from the Industrial Revolution up through today.

Why are we concerned with air chemistry? Why study air chemistry at all? In a general sense, consider that chemistry basically affects everything we do. Not a single moment of time goes by during which we are not affected in some way by a chemical substance, chemical process, or chemical reaction. Chemistry affects just about every area of our daily lives. In a specific sense, consider that almost every environmental and pollution problem we face today (and probably tomorrow) has a chemical basis. Without chemistry, it would be difficult, if not impossible, to study problems related to the quality of air that living organisms breathe, the nature and level of air pollutants, visibility, atmospheric esthetics, and climate, as well as those phenomena affecting the atmosphere (greenhouse effect, ozone depletion, groundwater contamination, toxic wastes, air pollution, and acid rain).

The task of the air science practitioner is to understand at least the most influential of the reactions linking atmospheric chemical constituents, as well as their principal sources and removal technologies. A basic review of chemistry fundamentals, chemical reactions in the atmosphere, and the chemical nature of atmospheric chemical species is given in this chapter.

Note that the air science practitioner is concerned in general with the atmospheric cycle—the same general phenomena that governs and produces all aspects of atmospheric chemistry. We will discuss air science as it relates to the unpolluted atmosphere, to highly polluted atmospheres, and to a wide range of gradations in between.

## 5.2 CHEMISTRY FUNDAMENTALS

*Note*: This text assumes that the air science student or interested reader who uses this book may or may not have some fundamental knowledge of chemistry. Thus, in this section, the topics have been selected with the goal of reviewing or providing only the essential chemical principles required to understand the nature of the air science problems we face and the chemistry involved with scientific and technological approaches to their solutions.

## 5.2.1 WHAT IS CHEMISTRY?

Chemistry is the science concerned with the composition of matter (gas, liquid, or solid) and of the changes that take place in it under certain conditions.

Every substance, material, and object in the environment is either a chemical substance or mixture of chemical substances. Your body is made up of chemicals, literally thousands of them. The food we eat, the clothes we wear, the fuel we burn, and the vitamins we take in from natural or synthetic sources are all products of chemistry, wrought either by the forces of nature or the hand of man. Chemistry is about matter—its constituents and consistency—and about measuring and quantifying matter.

What is matter? All matter can exist in three states: gas, liquid, or solid. Matter is composed of constantly moving minute particles termed molecules, which may be further divided into atoms.

Molecules that contain atoms of one kind only are known as *elements;* those that contain atoms of different kinds are called *compounds.*

Chemical compounds are produced by a chemical action that alters the arrangements of the atoms in the reacting molecules. Heat, light, vibration, catalytic action, radiation, or pressure, as well as moisture (for ionization), may be necessary to produce a chemical change. Examination and possible breakdown of compounds to determine their components is analysis, and the building up of compounds from their components is synthesis. When substances are brought together without changing their molecular structures, they are said to be *mixtures.*

*Organic substances* are compounds that contain carbon; all other substances are *inorganic substances.*

## 5.2.2 ELEMENTS AND COMPOUNDS

A pure substance is a material that has been separated from all other materials. Examples of such substances (indistinguishable from other pure samples of the same substance, no matter what procedures are used to purify them or what their origin is) are copper metal, aluminum metal, distilled water, table sugar, and oxygen. All samples of pure table sugar are alike and indistinguishable from all other samples.

Usually expressed in terms of percentage by mass, a substance is characterized as a material having a fixed composition. Distilled water, for example, is a pure substance consisting of approximately 11% hydrogen and 89% oxygen by mass. By contrast, a lump of coal is not a pure substance; it is a mixture, and its carbon content may vary from 35% to 90% by mass.

When substances cannot be broken down or decomposed into simpler forms of matter, they are called elements. The elements are the basic substances of which all matter is composed. At the present time, there are only 100+ known elements but well over a million known compounds. Of the 100+ elements, only 88 are present in detectable amounts on Earth, and many of these are rare. Table 5.1 shows that only 10 elements make up approximately 99% by mass of the Earth's crust, including the surface layer, the atmosphere, and the bodies of water. From Table 5.1, we see that the most abundant element on Earth is oxygen, which is found in the free state in the atmosphere, as well as in combined form with other elements in numerous minerals and ores. Table 5.1 also lists the symbols and atomic number of the 10 chemicals listed. The symbols consist of either one or two letters, with the first letter capitalized. The *atomic number* of an element is the number of protons in the nucleus.

## 5.2.3 CLASSIFICATION OF ELEMENTS

Each element may be classified as a metal, nonmetal, or metalloid. *Metals* are typically lustrous solids that are good conductors of heat and electricity; they melt and boil at high temperatures, possess relatively high densities, and are normally malleable (can be hammered into sheets) and ductile (can be drawn into a wire). Examples of metals are copper, iron, silver, and platinum. Almost all metals are solids (none is gaseous) at room temperature (mercury being the only exception).

Elements that do not possess the general physical properties just mentioned (i.e., they are poor conductors of heat and electricity, boil at relatively

TABLE 5.1. Elements Making up 99% of Earth's Crust, Oceans, and Atmosphere.

| Element | Symbol | Percent of Composition | Atomic Number |
|---------|--------|------------------------|---------------|
| Oxygen | O | 49.5% | 8 |
| Silicon | Si | 25.7% | 14 |
| Aluminum | Al | 7.5% | 13 |
| Iron | Fe | 4.7% | 26 |
| Calcium | Ca | 3.4% | 20 |
| Sodium | Na | 2.6% | 11 |
| Potassium | K | 2.4% | 19 |
| Magnesium | Mg | 1.9% | 12 |
| Hydrogen | H | 1.9% | 1 |
| Titanium | Ti | 0.58% | 22 |

low temperatures, do not possess a luster, and are less dense than metals) are called *nonmetals*. Most nonmetals at room temperature are either solids or gases (exception is bromine—a liquid). Nitrogen, oxygen, and fluorine are examples of gaseous nonmetals, while sulfur, carbon, and phosphorus are examples of solid nonmetals.

Several elements have properties resembling both metals and nonmetals. They are called *metalloids* (semimetals). The metalloids are boron, silicon, germanium, arsenic, tellurium, antimony, and polonium.

## 5.2.4 PHYSICAL AND CHEMICAL CHANGES

Internal linkages among a substance's units (between one atom or another) maintain its constant composition. These linkages are called chemical bonds. When a particular process occurs that involves the making and breaking of these bonds, we say that a chemical change or chemical reaction has occurred. In environmental science, combustion and corrosion are common examples of chemical changes that impact our environment.

Let's briefly consider a couple of chemical change examples. When a flame is brought into contact with a mixture of hydrogen and oxygen gases, a violent reaction takes place. The covalent bonds in the hydrogen ($H_2$) molecules and of the oxygen ($O_2$) molecules are broken, and new bonds are formed to produce molecules of water, $H_2O$.

The key point to remember is that whenever chemical bonds are broken or formed, or both, a chemical change takes place. The hydrogen and oxygen undergo a chemical change to produce water, a substance with new properties.

When mercuric oxide (a red powder) is heated, small globules of mercury are formed and oxygen gas is released. This mercuric oxide is changed chemically to form molecules of mercury and molecules of water.

By contrast, a physical change (nonmolecular change) is one in which the molecular structure of a substance is not altered. When a substance freezes, melts, or changes to vapor, the composition of each molecule does not change. For example, ice, steam, and liquid water all are made up of molecules containing two atoms of hydrogen and one atom of oxygen. A substance can be ripped or sawed into small pieces, ground into powder, or molded into a different shape without changing the molecules in any way.

The types of behavior that a substance exhibits when undergoing chemical changes are called its chemical properties. The characteristics that do not involve changes in the chemical identity of a substance are called its physical properties. All substances may be distinguished from one another by these properties, in much the same way as certain features (DNA, for example) distinguish one human being from another.

## 5.3  THE STRUCTURE OF THE ATOM

If a small piece of an element (say, copper) is divided and subdivided into the smallest piece possible, the result would be one particle of copper. This smallest particle of an element that is still representative of the element is called an *atom*.

Although infinitesimally small, the atom is composed of particles, principally electrons, protons, and neutrons. The most simple atom possible consists of a nucleus with a single *proton* (positively charged particle) and a single *electron* (negatively charged particle) traveling around it—an atom of hydrogen, which has an atomic weight of one because of the single proton. The atomic weight of an element is equal to the total number of protons and neutrons (neutral particles) in the nucleus of an atom of an element. Electrons and protons bear the same magnitude of charge but of opposite polarity.

The hydrogen atom has an atomic number of one because of its one proton. The *atomic number* of an element is equal to the number of protons in its nucleus. A neutral atom has the same number of protons and electrons. Therefore, in a neutral atom, the atomic number is also equal to the number of electrons in the atom. The number of neutrons in an atom is always equal to or greater than the number of protons, except in the atom of hydrogen.

The protons and neutrons of an atom reside in the nucleus. Electrons reside primarily in designated regions of space surrounding the nucleus, called *atomic orbitals* or *electron shells*. Only a prescribed number of electrons may reside in a given type of electron shell. With the exception of hydrogen (which has only one electron), two electrons are always close to the nucleus in an atom's innermost electron shell. In most atoms, other electrons are located in electron shells some distance from the nucleus.

While neutral atoms of the same element have an identical number of electrons and protons, they may differ by the number of neutrons in their nuclei. Atoms of the same element having different numbers of neutrons are called *isotopes* of that element.

### 5.3.1  PERIODIC CLASSIFICATION OF THE ELEMENTS

Through experience, scientists discovered that the chemical properties of the elements repeat themselves. Chemists summarize all such observations in the *periodic law:* the properties of the elements vary periodically with their atomic numbers.

In 1869, Dimitri Mendeleev, using relative atomic masses, developed the original form of what today is known as the *periodic table*—a chart of elements arranged in order of increasing proton number to show the similarities of chemical elements with related electronic configurations. The elements fall into vertical columns known as *groups*. Going down a group, the atoms of the elements all have the same outer shell structure but an increasing number of inner shells. Traditionally, the alkali metals are shown on the left of the table and the groups are numbered IA to VIIA, IB to VIIB, and 0 (for noble gases). It is now more common to classify all the elements in the middle of the table as transition elements and to regard the nontransition elements as main group elements numbered from I to VII, with the noble gases in group 0. Horizontal rows in the table are *periods*. The first three are called short periods; the next four (which include transition elements) are long periods. Within a period, the atoms of all the elements have the same number of shells but with a steadily increasing number of electrons in the outer shell.

The periodic table is an important tool for learning chemistry because it tabulates a variety of information in one spot. For example, we can immediately determine the atomic number of the elements because they are tabulated on the periodic table (see Figure 5.1). We can also readily identify which elements are metals, nonmetals, and metalloids. Usually a bold zigzag line separates metals from nonmetals, while those elements lying to each immediate side of the line are metalloids. Metals fall to the left of the line, and nonmetals fall to the right of it.

### 5.3.2 MOLECULES AND IONS

When elements other than noble gases (which exist as single atoms) exist in either the gaseous or liquid state of matter at room conditions, they consist of units containing pairs of like atoms. These units are called molecules. For example, we generally encounter oxygen, hydrogen, chlorine, and nitrogen as gases. Each exists as a molecule having two atoms. These molecules are symbolized by the notations $O_2$, $H_2$, $Cl_2$, and $N_2$, respectively.

The smallest particle of many compounds is also the molecule. Molecules of compounds contain atoms of two or more elements. Water molecules, for example, consist of two atoms of hydrogen and one atom of oxygen ($H_2O$). Methane molecules consist of one carbon atom and four hydrogen atoms ($CH_4$).

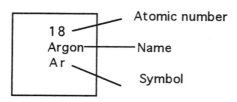

**Figure 5.1** The element argon as it is commonly shown in one of the horizontal boxes in the periodic table.

Not all compounds occur naturally as molecules. Many of them occur as aggregates of oppositely charged atoms (or groups of atoms) called *ions*. Atoms become charged by gaining or losing some of their electrons. Atoms of metals, for example, that lose their electrons become positively charged, and atoms of nonmetals that gain electrons become negatively charged.

### 5.3.3 CHEMICAL BONDING

When compounds form, the atoms of one element become attached to or associated with atoms of other elements by forces called chemical bonds. Chemical bonding is a strong force of attraction holding atoms together in a molecule. There are various types of chemical bonds. *Ionic* bonds can be formed by transfer of electrons. For instance, the calcium atom has an electron configuration of two electrons in its outer shell. The chlorine atom has seven outer electrons. If the calcium atom transfers two electrons (one to each chlorine atom), it becomes a calcium ion with the stable configuration of an inert gas. At the same time, each chlorine (having gained one electron) becomes a chlorine ion, also with an inert gas configuration. The bonding in calcium chloride is the electrostatic attraction between the ions.

*Covalent* bonds are formed by the sharing of *valence* (the number of electrons an atom can give up or acquire to achieve a filled outer shell) electrons. Hydrogen atoms, for instance, have one outer electron. In the hydrogen molecule, $H_2$, each atom contributes one electron to the bond. Consequently, each hydrogen atom has control of two electrons—one of its own and the second from the other atom—giving it the electron configuration of an inert gas. In the water molecule ($H_2O$), the oxygen atom (with six outer electrons) gains control of an extra two electrons supplied by the two hydrogen atoms. Similarly, each hydrogen atom gains control of an extra electron from the oxygen.

Chemical compounds are often classified into one of two groups, based on the nature of the bonding between their atoms. As you might expect, chemical compounds consisting of atoms bonded together by means of ionic bonds are called ionic compounds. Compounds whose atoms are bonded together by covalent bonds are called covalent compounds.

Most ionic and covalent compounds provide some interesting contrasts. For example, ionic compounds have higher melting points, boiling points, and solubility in water than covalent compounds. Ionic compounds are nonflammable while covalent compounds are flammable. Ionic compounds molten in water solutions conduct electricity. Molten covalent compounds do not conduct electricity. Ionic compounds generally exist as solids at room temperature, while covalent compounds can exist as gases, liquids, or solids at room temperature.

## 5.3.4 CHEMICAL FORMULAS AND EQUATIONS

Chemists have developed a shorthand method of writing chemical formulas. Elements are represented by groups of symbols called formulas. A common compound is sulfuric acid; its formula is $H_2SO_4$. The formula indicates that the acid is composed of two atoms of hydrogen, one atom of sulfur, and four atoms of oxygen. However, this is not a recipe for making the acid. The formula does not tell you how to prepare the acid, only what is in the acid.

A *chemical equation* tells what elements and compounds are present before and after a chemical reaction. Sulfuric acid poured over zinc will cause the release of hydrogen and the formation of zinc sulfate. This is shown by the following equation:

$$Zn + H_2SO_4 \rightarrow ZnSO_4 + H_2$$

(zinc)  (sulfuric acid)    (zinc sulfate)  (hydrogen)

One atom (also one molecule) of zinc unites with one molecule of sulfuric acid, giving one molecule of zinc sulfate and one molecule (two atoms) of hydrogen. Notice the same number of atoms of each element is on each side of the arrow. However, the atoms are combined differently.

## 5.3.5 MOLECULAR WEIGHTS, FORMULAS, AND THE MOLE

The relative weight of a compound that occurs as molecules is called the *molecular weight*—the sum of the atomic weights of each atom that comprises the molecule. Consider the water molecule. Its molecular weight is determined as follows:

$$2 \text{ hydrogen atoms} = 2 \times 1.008 = 2.016$$
$$\underline{1 \text{ oxygen atom} = 1 \times 15.999 = 15.999}$$
$$\text{molecular weight of } H_2O = 18.015$$

Thus, the molecular weight of a molecule is simply the sum of the atomic weights of all the constituent atoms. If we divide the mass of a substance by its molecular weight, the result is the mass expressed in *moles (mol)*. Usually, the mass is expressed in grams, in which case the moles are written as g-moles; similarly, if mass is expressed in pounds, the result would be lb-moles. One g-mole contains $6.022 \times 10^{23}$ molecules (*Avogadro's number*, in honor of the scientist who first suggested its existence), and one lb-mole contains about $2.7 \times 10^{26}$ molecules.

$$\text{moles} = \frac{\text{mass}}{\text{molecular weight}} \qquad (5.1)$$

The relative weight of a compound that occurs as formula units is called the *formula weight*—the sum of the atomic weights of all atoms that comprise one formula unit. Consider sodium fluoride. Its formula weight is determined as follows:

$$1 \text{ sodium ion} = 22.990$$

$$\underline{1 \text{ flouride ion} = 18.998}$$

$$\text{formula weight of NaF} = 41.988$$

## 5.4 PHYSICAL AND CHEMICAL PROPERTIES OF MATTER

There are two basic types of properties (characteristics) of matter: physical and chemical. *Physical properties* of matter are those that do not involve a change in the chemical composition of the substance. Among these properties are hardness, color, boiling point, electrical conductivity, thermal conductivity, specific heat, density, solubility, and melting point. These properties may change with a change in temperature or pressure. Those changes that do not alter the chemical composition of the substance are called physical changes. When heat is applied to solid ice to convert it to liquid water, no new substance is produced, but the appearance has changed; melting is a physical change. Other examples of physical changes are dissolving sugar in water, heating a piece of metal, and evaporating water.

The physical properties most commonly used in describing and identifying particular kinds of matter are density, color, and solubility. *Density* $(d)$ is mass per unit volume and is expressed by the equation

$$d = \frac{\text{mass}}{\text{volume}} \qquad (5.2)$$

All matter has weight and takes up space; it also has density, which depends on weight and space. We commonly say that a certain material will not float in water because it is heavier than water. What we really mean is that a particular material is more dense than water. The density of an element differs from the density of any other element. The densities of liquids and solids are normally given in units of grams per cubic centimeter ($g/cm^3$), which is the same as grams per milliliter (g/mL). The advantage of using the physical property color is that no chemical or physical tests are required. *Solubility* refers to the degree to which a substance dissolves in a liquid such as water. In air science, the density, color, and solubility of a substance are important physical properties that aid in the determination of various pollutants, stages of pollution or treatment, the remedial actions required to clean up volatile toxic/hazardous waste spills, and other environmental problems.

The properties involved in the transformation of one substance into another are known as *chemical properties*. For example, when a piece of wood burns, oxygen in the air unites with the different substances in the wood to form new substances. Another example is corroded iron. During the iron corrosion process, oxygen combines with the iron and water to form a new substance commonly known as rust. Changes that result in the formation of new substances are known as chemical changes.

## 5.4.1 STATES OF MATTER

As pointed out earlier, the three common states (or phases) of matter are the solid state, the liquid state, and the gaseous state. In the solid state, the molecules or atoms are in a relatively fixed position. The molecules are vibrating rapidly but about a fixed point. Because of this definite position of the molecules, a solid holds its shape. A solid occupies a definite amount of space and has a fixed shape.

When the temperature of a gas is lowered, the molecules of the gas slow down. If the gas is cooled sufficiently, the molecules slow down so much that they lose the energy needed to move rapidly throughout their container. The gas may turn into a liquid. Common liquids are water, oil, and gasoline. A *liquid* is a material that occupies a definite amount of space but takes the shape of the container.

In some materials, the atoms or molecules have no special arrangement at all. Such materials are called gases. Oxygen, carbon dioxide, and nitrogen are common gases. A *gas* is a material that takes the exact volume and shape of its container.

Although the three states of matter discussed here are familiar to most people, the change from one state to another is of primary interest to environmentalists. Changes in matter that include water vapor changing from the gaseous state to liquid precipitation or a spilled liquid chemical changed to a semisolid substance (by addition of chemicals that aid the cleanup effort) are just two ways that changing from one state to another can have an impact on environmental concerns.

## 5.4.2 THE GAS LAWS

The gas laws were explained earlier in the text; for the purpose of review and because of their importance to this text, they will be explained again briefly.

The atmosphere is composed of a mixture of gases, the most abundant of which are nitrogen, oxygen, argon, carbon dioxide, and water vapor. The *pressure* of a gas is the force that the moving gas molecules exert upon a unit area. A common unit of pressure is newton per square meter, $N/m^2$, called a pascal (Pa). An important relationship exists among the pressure, volume, and temperature of a gas. This relation is known as the *Ideal Gas Law* and can be stated as

$$\frac{P_1 V_1}{T_1} = \frac{P_2 V_2}{T_2} \tag{5.3}$$

where $P_1$, $V_1$, $T_1$ are pressure, volume, and absolute temperature at time 1, and $P_2$, $V_2$, $T_2$, are pressure, volume, and absolute temperature at time 2. A gas is called perfect (or ideal) when it obeys this law.

A temperature of 0°C (273 K) and a pressure of 1 atmosphere (atm) have been chosen as standard temperature and pressure (STP). At STP the volume of 1 mole of ideal gas is 22.4 L.

### 5.4.3 LIQUIDS AND SOLUTIONS

Solutions, which are homogenous mixtures, can be solid, gaseous, or liquid. However, the most common solutions are liquids. The substance in excess in a solution is called the solvent. The substance dissolved is the solute. Solutions in which water is the solvent are called aqueous solutions. A solution in which the solute is present in only a small amount is called a dilute solution. If the solute is present in large amounts, the solution is a concentrated solution. When the maximum amount of solute possible is dissolved in the solvent, the solution is called a saturated solution.

The concentration (or amount of solute dissolved) is frequently expressed in terms of the molar concentration. The *molar concentration, or molarity,* is the number of moles of solute per liter of solution. Thus, a one molar solution (written 1.0 M) has 1 g formula weight of solute dissolved in 1 L of solution. In general,

$$\text{molarity} = \frac{\text{moles of solute}}{\text{number of liters of solution}} \tag{5.4}$$

Note that the number of liters of solution (not the number of liters of solvent) is used.

### *Example 24*

Exactly 40 g of sodium chloride (NaCl) (table salt) were dissolved in water, and the solution was made up to a volume 0.80 L of solution. What was the molar concentration (M) of sodium chloride in the resulting solution?

*Solution:* First find the number of moles of salt.

$$\text{number of moles} = \frac{40\,\text{g}}{58.5\,\text{g}/\text{mole}} = 0.68\,\text{mole}$$

$$\text{molarity} = \frac{0.68\,\text{mole}}{0.80\,\text{L}} = 0.85\,\text{M}$$

## 5.4.4 THERMAL PROPERTIES

Knowing the thermal properties of chemicals and other substances is important to the air science practitioner. Such knowledge is used in hazardous materials spill mitigation and in solving many other complex environmental problems. Heat is a form of energy. Whenever work is performed, usually a substantial amount of heating is caused by friction. The conservation of energy law tells us that the work done plus the heat energy produced must equal the original amount of energy available; that is,

$$\text{total energy} = \text{work done} + \text{heat produced} \qquad (5.5)$$

As air or environmental scientists, technicians, and/or practitioners, we are concerned with several properties related to heat for particular substances. Those thermal properties that we need to be familiar with are discussed in the following sections.

A traditional unit for measuring heat energy is the calorie. A *calorie* (cal) is defined as the amount of heat necessary to raise 1 g of pure liquid water by 1°C at normal atmospheric pressure. In SI units, 1 cal is 4.186 J (Joule).

Do not confuse the calorie just defined with the one used when discussing diets and nutrition. A kilocalorie is 1000 calories—the amount of heat necessary to raise the temperature of 1 kg of water by 1°C.

In the British system of units, the unit of heat is the British thermal unit, or Btu. One *Btu* is the amount of heat required to raise 1 lb of water 1°F at normal atmospheric pressure (1 atm).

### 5.4.4.1 Specific Heat

Earlier, it was stated that 1 kcal of heat is necessary to raise the temperature of 1 kg of water 1°C. Other substances require different amounts of heat to raise the temperature of 1 kg of the substance one degree. The specific heat of a substance is the amount of heat in kilocalories necessary to raise the temperature of 1 kg of the substance 1°C.

The units of specific heat are kcal/kg°C or, in SI units, J/kg°C. The specific heat of pure water, for example, is 1.000 kcal/kg°C. This is 4186 J/kg°C.

The greater the specific heat of a material, the more heat is required. The greater the mass of the material or the greater the temperature change desired, the more the heat is required.

The amount of heat necessary to change 1 kg of a solid into a liquid at the same temperature is called the *latent heat of fusion* of the substance. The temperature of the substance at which this change from solid to liquid takes place is known as the *melting point.* The amount of heat necessary to change 1 kg of a liquid into a gas is called the *latent heat of vaporization*. When this point has been reached, the substance is all in the gas state. The temperature of the substance at which this change from liquid to gas occurs is known as the *boiling point.*

## 5.5  ACIDS + BASES → SALTS

When acids and bases are combined in the proper proportions they neutralize each other, each losing its characteristic properties and forming a salt and water. For example,

$$NaOH + HCl \rightarrow NaCl + H_2O$$

which is

sodium hydroxide + hydrochloric acid → sodium chloride + water

The acid–base–salt concept originated with the beginning of chemistry and is very important in the environment, the life processes, and in industrial chemistry.

The word *acid* is derived from the Latin *acidus,* which means sour. The sour taste is one of the properties of acids (however, the student should never actually taste an acid in the laboratory or anywhere else). An *acid* is a substance that, in water, produces hydrogen ions, $H^+$, and has the following properties:

(1) Conducts electricity
(2) Tastes sour
(3) Changes the color of blue litmus paper to red
(4) Reacts with a base to neutralize its properties
(5) Reacts with metals to liberate hydrogen gas

A *base* is a substance that produces hydroxide ions, OH⁻, and/or accepts H⁺, and when dissolved in water it has the following properties:

(1) Conducts electricity
(2) Changes the color of red litmus paper to blue
(3) Tastes bitter and feels slippery
(4) Reacts with an acid to neutralize the acid's properties

### 5.5.1 pH SCALE

A common way to determine whether a solution is an acid or a base is to measure the concentration of hydrogen ions (H⁺) in the solution. The concentration can be expressed in powers of 10 but is more conveniently expressed as *pH* (the "p" is from the German word *poentz* and the "H" stands for hydrogen). For example, pure water has $1 \times 10^{-7}$ grams of hydrogen ions per liter. The negative exponent of the hydrogen ion concentration is called the pH of the solution. The pH of water is 7—a neutral solution. A concentration of $1 \times 10^{-12}$ has a pH of 12. A pH less than 7 indicates an acid solution, and a pH greater than 7 indicates a basic solution (see Table 5.2).

The pH of substances found in our environment varies in value. Acid–base reactions are among the most important in environmental science. In diagnosing various environmental problems such as acid rain problems, hazardous materials spills into lakes or streams, and the effect of point or nonpoint source pollution into our streams, rivers, lakes, and ponds, pH value is important. Along with diagnosis, remediation and/or prevention are also important. For example, to protect a local ecosystem, volatile wastes often

TABLE 5.2. Standard pH Scale.

| pH | Concentration of H⁻ Ions | Acidic/Basic |
|----|--------------------------|--------------|
| 1 | $1.0 \times 10^{-1}$ mole/liter | Very acidic |
| 2 | $1.0 \times 10^{-2}$ mole/liter | |
| 3 | $1.0 \times 10^{-3}$ mole/liter | |
| 4 | $1.0 \times 10^{-4}$ mole/liter | |
| 5 | $1.0 \times 10^{-5}$ mole/liter | |
| 6 | $1.0 \times 10^{-6}$ mole/liter | Slightly acidic |
| 7 | $1.0 \times 10^{-7}$ mole/liter | Neutral |
| 8 | $1.0 \times 10^{-8}$ mole/liter | Slightly basic |
| 9 | $1.0 \times 10^{-9}$ mole/liter | |
| 10 | $1.0 \times 10^{-10}$ mole/liter | |
| 11 | $1.0 \times 10^{-11}$ mole/liter | |
| 12 | $1.0 \times 10^{-12}$ mole/liter | |
| 13 | $1.0 \times 10^{-13}$ mole/liter | |
| 14 | $1.0 \times 10^{-14}$ mole/liter | Very basic |

TABLE 5.3. pH of Common Substances.

| Substance | | pH |
|---|---|---|
| Battery acid | | 0.0 |
| Gastric juice | | 1.2 |
| Lemons | | 2.3 |
| Vinegar | | 2.8 |
| Soft drinks | | 3.0 |
| Apples | | 3.1 |
| Grapefruit | | 3.1 |
| Wine | | 3.2 |
| Oranges | | 3.5 |
| Tomatoes | | 4.2 |
| Beer | | 4.5 |
| Bananas | | 4.6 |
| Carrots | | 5.0 |
| Potatoes | | 5.8 |
| Coffee | | 6.0 |
| Milk (cow) | | 6.5 |
| Pure water | Neutral | 7.0 |
| Blood (human) | | 7.4 |
| Eggs | | 7.8 |
| Sea water | | 8.5 |
| Milk of magnesia | | 10.5 |
| Oven cleaner | | 13.0 |

require neutralization before being released into the environment. Table 5.3 gives an approximate pH of some common substances.

## 5.6 ORGANIC CHEMISTRY

Organic chemistry is the branch of chemistry concerned with compounds of carbon. The science of organic chemistry is incredibly complex and varied. There are millions of different organic compounds known today, and 100,000+ of these are products of synthesis, unknown in nature. The next few pages, will provide a very "basic" introduction to some of the most common organic substances important to environmental science (important because of their toxicities as pollutants and other hazards) so that they will be less foreign when we encounter them later in the text.

Before 1828, scientists thought that organic compounds could only be made by plants and animals (living things). In that year, Friedrich Wohler made urea from ammonium cyanate, thus disproving the theory that urea could only be made from living things. Because of his discovery, the science of organic chemistry was born.

Organic compounds are components of all the familiar commodities our technological world requires. Examples of such commodities are motor and heating fuels, adhesives, cleaning solvents, paints, varnishes, plastics, refrigerants, aerosols, textiles, fibers, and resins, among many others.

From an environmental science perspective, the principal concern about organic compounds is that they are pollutants of water, air, and soil environments. As such, they are safety and health hazards. They are also combustible or flammable substances, with few exceptions. From a health standpoint, they have the ability to cause a wide range of detrimental health effects. In humans, some of these compounds damage the kidneys, liver, and heart; others depress the central nervous system; and several are suspected of causing cancer. If human beings are subject to such health hazards from organic compounds, the logical question for environmental scientists to ask is, "What about their impact on delicate ecosystems?"

## 5.6.1 ORGANIC COMPOUNDS

The molecules of organic compounds have one feature in common: one or more carbon atoms that covalently bond to other atoms—that is, pairs of electrons are shared between atoms. A carbon atom may share electrons with other nonmetallic atoms and also with other carbon atoms. As Figure 5.2 shows, methane, carbon tetrachloride, and carbon monoxide are compounds having molecules in which the carbon atom is bonded to other nonmetallic atoms.

When carbon atoms share electrons with other carbon atoms, we find that two carbon atoms may share electrons in such a manner that they form one of the following: carbon–carbon single bonds (C–C); carbon–carbon double bonds (C=C); or carbon–carbon triple bonds (C≡C). Each bond written here as a dash (–) is a shared pair of electrons. Figure 5.3 illustrates the bonding of molecules of ethane, ethylene, and acetylene—compounds having molecules with only two carbon atoms. Molecules of ethane possess carbon–carbon single bonds; molecules of ethylene possess carbon–carbon double bonds; and molecules of acetylene possess carbon–carbon triple bonds.

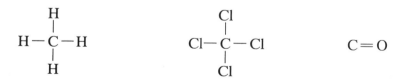

Methane          Carbon tetrachloride    Carbon monoxide

**Figure 5.2** Carbon atoms sharing their electrons with the electrons of other nonmetallic atoms such as hydrogen, chlorine, and oxygen and their resulting compounds.

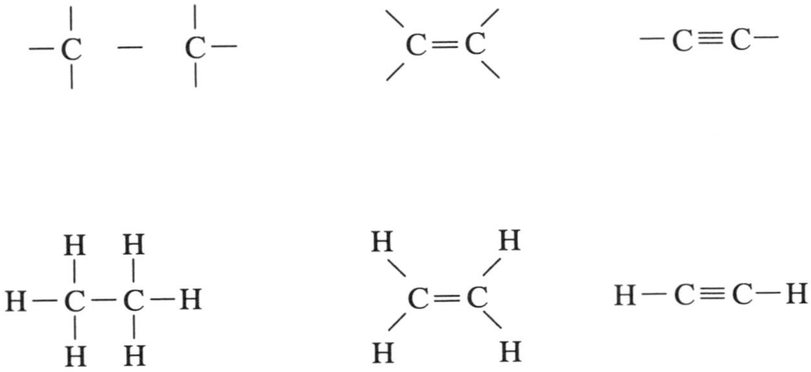

**Figure 5.3** (top) Two carbon atoms may share their own electrons in either of three ways. (bottom) When they further bond to hydrogen atoms, the resulting compounds are ethane, ethene, and acetylene, respectively.

Covalent bonds between carbon atoms in molecules of more complex organic compounds may be linked into chains, including branched chains, or into rings.

### 5.6.2 HYDROCARBONS

The simplest organic compounds are the *hydrocarbons,* compounds whose molecules are composed only of carbon and hydrogen atoms. All hydrocarbons are broadly divided into two groups: aliphatic and aromatic hydrocarbons.

### 5.6.3 ALIPHATIC HYDROCARBONS

*Aliphatic hydrocarbons* can be characterized by the chain arrangements of their constituent carbon atoms. They are divided into the alkanes, alkenes, and alkynes.

The *alkanes* (also called paraffins or aliphatic hydrocarbons) are saturated hydrocarbons (hydrogen content is at its maximum) with the general formula $C_nH_{2n+2}$. In systematic chemical nomenclature, alkane names end in the suffix *-ane.* They form the alkane series methane ($CH_4$), ethane ($C_2H_6$), propane ($C_3H_8$), butane ($C_4H_{10}$), etc. The lower members of the series are gases; the high-molecular-weight alkanes are waxy solids. Alkanes are present in natural gas and petroleum.

*Alkenes* (olefins) are unsaturated hydrocarbons (can take on hydrogen atoms to form saturated hydrocarbons) containing one or more double carbon–carbon bonds in their molecules. In systematic chemical nomenclature, alkene names end in the suffix *-ene.* Alkenes with only one double bond form the alkene series starting with ethene (the gas that is liberated when food rots); ethylene, $CH_2:CH_2$, propene; $CH_3CH:CH_2$; etc.

Alkynes (acetylenes) are unsaturated hydrocarbons that contain one or more triple carbon–carbon bonds in their molecules. In systematic chemical nomenclature, alkyne names end in the suffix *-yne,* such as acetylene, $H–C\equiv C–H$.

### 5.6.4  AROMATIC HYDROCARBONS

*Aromatic hydrocarbons* are those unsaturated organic compounds that contain a benzene ring in their molecules or that have chemical properties similar to benzene, a clear, colorless, water-insoluble liquid. Benzene rapidly vaporizes at room temperature, and the molecular formula is $C_6H_6$. The molecular structure of benzene is commonly represented by a hexagon with a circle inside as shown below:

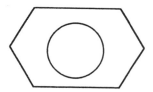

### 5.7  ENVIRONMENTAL CHEMISTRY

Environmental chemistry is a blend of aquatic, atmospheric, and soil chemistry and the "chemistry" generated by human activities in those areas. The focus of this text, of course, deals with the environmental medium air. The environmental effects brought about by human influence cannot be ignored.

At this point, we must understand that the chemistry that makes up the air medium—as well as the chemical reactions that take place to preserve and to destroy this medium—are vital information in our study of environmental science. As we proceed through our discussion of air, we will be concerned with the environmental impact of human activities—mining, acid rain, erosion from poor cultivation practices, disposal of volatile hazardous wastes, photochemical reactions (smog), air pollutants such as particulate matter, the greenhouse effect, ozone, and water degradation problems related to organic, inorganic, and biological pollutants. All of these activities and problems have something to do with chemistry, and the remediation or mitigation processes to solve them are also tied to chemistry.

Having said this, we turn to a discussion of atmospheric chemistry—the real core of air chemistry. To say that environmental science, environmental studies, and environmental engineering related to air science are built upon a strong foundation of chemistry is to mildly understate chemistry's real importance and relevance.

## 5.8 ATMOSPHERIC CHEMISTRY AND RELATED PHENOMENA

*Note*: In the brief discussion that follows describing the chemistry of Earth's atmosphere, keep in mind that the atmosphere as it is at present (during the age of Man) is what is referred to. The atmosphere previous to this period was chemically quite different (see Chapter 6). Note also that "atmospheric chemistry" is a scientific discipline that can stand on its own. The "nuts and bolts" of atmospheric chemistry are beyond the scope of this book. Here, certain important atmospheric chemistry phenomena are highlighted.

You should recall that the full range of chemistry (including slow and fast reactions, dissolving crystals, and precipitation of solids) all occur in the atmosphere, and that the atmosphere is as Graedel and Crutzen (1995) described it—a "flask without walls." We also pointed out that (excluding highly variable amounts of water vapor) more than 99% of the molecules constituting Earth's atmosphere are nitrogen, oxygen, and chemically inert gases (noble gases such as argon, etc.).

The chemistry (and thus the reactivity) of these natural gases (nitrogen, oxygen, carbon dioxide, argon, and others) is well known. The other reactive chemicals (anthropogenically produced) that are part of Earth's atmosphere are also known, but there are still differing opinions on their "exact" total effect on our environment. For example, methane is by far the most abundant reactive compound in the atmosphere, and it currently is at a ground-level concentration (in the northern hemisphere) of about 1.7 ppmv. We know significant amounts of information about methane (its generation and fate when discharged) and its influence on the atmosphere; however, we are still conducting research to find out more—as we should.

Many different reactive molecules (other than methane) exist in the atmosphere. We may not be familiar with each of these reactants, but many of us certainly are familiar with their consequences: the ozone hole, the greenhouse effect and global warming, smog, acid rain, the rising tide, and so on. It may surprise you to know, however, that the total amount of all these reactants in the atmosphere is seldom more than 10 ppmv anywhere in the world at any given time. The significance should be obvious: the atmospheric problems currently occurring on Earth are the result of less than one thousandth of 1% of all of the molecules in the atmosphere. This indicates that environmental damage (causing global atmospheric problems) can result from far less than the tremendous amounts of reactive substances we might imagine are dangerous.

The quality of the air we breathe, the visibility and atmospheric esthetics, and our climate (all of which are dependent upon chemical phenomena that occur in the atmosphere) are important to our health and to our quality of life. Global atmospheric problems, such as the nature and level of air pollutants, concern the air science practitioner the most because they affect our health and our quality of life.

Let's take a look at some of the important chemical species and their reactions within (primarily) the stratosphere of our atmosphere.

## 5.8.1 PHOTOCHEMICAL REACTION—SMOG PRODUCTION

A *photochemical reaction*, generally, is any chemical reaction in which light is produced or light initiates the reaction. Light can initiate reactions by exciting atoms or molecules and making them more reactive. The light energy becomes converted to chemical energy.

The photochemical reaction we are concerned with here is the absorption of electromagnetic solar radiation (light) by chemical species, which causes the reactions. The ability of electromagnetic radiation to cause photochemical reactions to occur is a function shown in the relationship

$$E = h\nu \tag{5.6}$$

where

$E$ = energy of a photon
$\nu$ = frequency
$h$ = Planck's constant, $6.62 \times 10^{-27}$

The major photochemical reaction we are concerned with is the one that produces photochemical smog. *Photochemical smog* is initiated by nitrogen dioxide, which absorbs the visible or ultraviolet energy of sunlight, forming nitric oxide to free atoms of oxygen (O), which then combine with molecular oxygen ($O_2$) to form *ozone* ($O_3$). In the presence of hydrocarbons (other than methane) and certain other organic compounds, a variety of chemical reactions takes place. Some 80 separate reactions have been identified or postulated.

The photochemical reaction that produces the smog we are familiar with is dependent on two factors:

(1) Smog concentration is linked to both the amount of sunlight and hydrocarbons present.

(2) The amount is dependent on the initial concentration of nitrogen oxides.

In the production of photochemical smog, many different substances are formed in sequence, including acrolein, formaldehyde, and PAN (peroxyacetylnitrate). Photochemical smog (the characteristic haze of minute droplets) is a result of condensed, low-volatility organic compounds. The organics irritate the eyes and also, together with ozone, can cause severe damage to leafy plants. Photochemical smog tends to be most intense in the early afternoon when sunlight intensity is greatest. It differs from traditional smog (Los Angeles-type smog), which is most intense in the early morning and is dispersed by solar radiation.

In addition to the photochemical reactions that produce smog, many other chemical reactions take place in Earth's atmosphere, including ozone production, production of free radicals, chain reactions, oxidation processes, acid–base reactions, the presence of electrons and positive ions, and many others.

### 5.9 SUMMARY

To study these chemical phenomena and to gain understanding of the effects of electromagnetic radiation from the sun on the atmosphere; the interchange of chemical species with particles; the absorption of solar radiation by air molecules; the excited, energetic, and reactive species produced by absorption of light; and the interchange of molecular species and particles between the atmosphere and Earth's surface, we must be well versed in and have a fundamental understanding of basic chemical concepts. Of course, an air scientist who must solve environmental problems and understand environmental remediation cleanup processes, such as emission control systems, must be well grounded in chemical principles and the techniques of chemistry in general, because many of them are being used today to solve these problems.

### 5.10 REFERENCES

Graedel, T. E. & Crutzen, P. J., *Atmosphere, Climate, and Change*. New York: Scientific American, 1995.

Meyer, E., *Chemistry of Hazardous Materials*, 2nd edition. Englewood Cliffs, NJ: Prentice Hall, 1989.

# PART II

# Atmospheric Science

# CHAPTER 6

# The Atmosphere

*"This most excellent canopy, the air, look you, this brave o'erhanging firmament, this majestical roof fretted with golden fire . . ." (Shakespeare, Hamlet)*

## 6.1 INTRODUCTION

SEVERAL theories of cosmogony attempt to explain the origin of the universe. Without speculating on the validity of any one theory, the following is simply the author's view.

The time: 4,500 million years ago.

The vast void held only darkness—everywhere. Light had no existence.

After eons of time, darkness came to a sudden, smashing, shattering, annihilating, cataclysmic end—and there was light. This new force replaced darkness, lighting up the expanse without end, creating a brightness fed by billions of glowing round masses.

With the light was heat energy, which shone and warmed and transformed nothingness into mega-mega-mega trillions of super-excited ions, molecules, and atoms. Heat of unimaginable proportions formed gases—gases we don't even know how to describe or how to quantify, but they were everywhere.

With light, energy, heat, and gases present, the stage was set for the greatest show of all time—the formation of the Universe.

Over time—time in stretches so vast we cannot contemplate them meaningfully—heat, light, energy, and gases all came together and grew, like an expanding balloon, into one solid growing mass, until it had reached the point of no return—explosion level—the biggest bang of all time.

The Big Bang sent masses of hot gases in all directions, to the farthest reaches of the vast, wide, measureless void. Clinging together as they rocketed, soared, and swirled, forming galaxies that gradually settled into their slow arcs through the void, constantly propelled away from their origin,

**113**

these masses began their eternal evolution. Two masses concern us—the Sun and Earth.

Forces well beyond the power of the Sun (beyond anything imaginable) stationed our massive gaseous orb approximately 93,000,000 miles from the dense molten core (then enveloped in cosmic gases and the dust of time) that eventually became the insignificant mass (in universal scale) we now call Earth.

Distant from the Sun, Earth's mass began to cool slowly, more slowly than we can imagine. While the dust and gases cooled, Earth's inner core, mantle, and crust began to form, which was no more a quiet or calm evolution than the revolution that cast it into the void had been.

This transformation was downright violent: the cooling surface only a facade for the internal machinations going on inside. Out-gassing from huge, deep destructive vents (we would call them volcanoes today) and continuous eruptions delivered two main ingredients: magma and gas.

The magma worked to form the primitive features of Earth's early crust. The gases worked to form Earth's initial atmosphere—our point of interest.

About 4 billion years before the present, Earth's early atmosphere was chemically reducing, consisting primarily of methane, ammonia, water vapor, and hydrogen. For life as we know it today, it was inhospitable.

Earth's initial atmosphere was not a calm, quiescent environment. On the contrary, it was an environment best characterized as dynamic, ever changing, with bombardment after bombardment of intense, bond-breaking ultraviolet light. Intense lightning and radiation from radionuclides also provided energy to bring about chemical reactions that resulted in the production of relatively complicated molecules, including amino acids and sugars (the building blocks of life).

About 3.5 billion years before the present, primitive life formed in two radically different theaters: on Earth's shifting, solid ground and below the primordial seas near hydrothermal vents that spotted the unsteady sea floor.

Initially, on Earth's unstable surface these very primitive lifeforms derived their energy from fermentation of organic matter formed by chemical and photochemical processes; then they gained the ability to produce organic matter ($CH_2O$) by photosynthesis. Thus, the stage was set for the massive biochemical transformation that resulted in the production of almost all the atmosphere's $O_2$.

Initially, the $O_2$ produced was quite toxic to primitive lifeforms. However, much of this oxygen was converted to iron oxides by reaction with soluble iron. This process formed enormous deposits of iron oxides, the existence of which provides convincing evidence for the liberation of $O_2$ in the primitive atmosphere.

Eventually, enzyme systems developed that enabled organisms to mediate the reaction of waste-product oxygen with oxidizable organic matter in the sea. Later, the mode of waste gradient disposal was used by organisms to produce energy by respiration, the mechanism by which nonphotosynthetic organisms still obtain energy. In time, $O_2$ accumulated in the atmosphere. In addition to providing an abundant source of oxygen for respiration, the accumulated atmospheric oxygen formed an ozone ($O_3$) shield, which absorbs bond-rupturing ultraviolet radiation.

With the $O_3$ shield protecting tissue from destruction by high-energy ultraviolet radiation, the Earth, although still hostile to lifeforms we are familiar with, became a much more hospitable environment for life (self-replacing molecules), and gradually, lifeforms were able to move from the sea (where they flourished next to the hydrothermal gas vents) to the land. And from that point to the present, Earth's atmosphere became more lifeform friendly.

## 6.2 EARTH'S THIN SKIN

Shakespeare likened it to a majestic overhanging roof; others have likened it to the skin of an apple. Both of these descriptions of our atmosphere are fitting, as is it being described as the Earth's envelope, veil, or gaseous shroud. The atmosphere is more like the apple skin, however. This thin skin, or layer, contains the life-sustaining oxygen (21%) required by all humans and many other lifeforms; the carbon dioxide (0.03%) so essential for plant growth; the nitrogen (78%) needed for chemical conversion to plant nutrients; the trace gases such as methane, argon, helium, krypton, neon, xenon, ozone, and hydrogen; and varying amounts of water vapor and airborne particulate matter.

Gravity holds about half the weight of a fairly uniform mixture of these gases in the lower 18,000 feet of the atmosphere; approximately 98% of the material in the atmosphere is below 100,000 feet.

Atmospheric pressure varies from 1000 millibars (mb) at sea level to 10 mb at 100,000 feet. From 100,000 to 200,000 feet the pressure drops from 9.9 mb to 0.1 mb.

From Figure 6.1 we see that the atmosphere has a thickness of 40–50 miles; however, here we are primarily concerned with the troposphere, the part of the Earth's atmosphere that extends from the surface to a height of about 27,000 feet (see Figure 6.1) above the poles, about 36,000 feet in mid-latitudes, and about 53,000 ft over the equator. Above the troposphere is the stratosphere, a region that increases in temperature with altitude (the warming is caused by absorption of the sun's radiation by ozone) until it reaches its upper limit at 260,000 feet (see Figure 6.1).

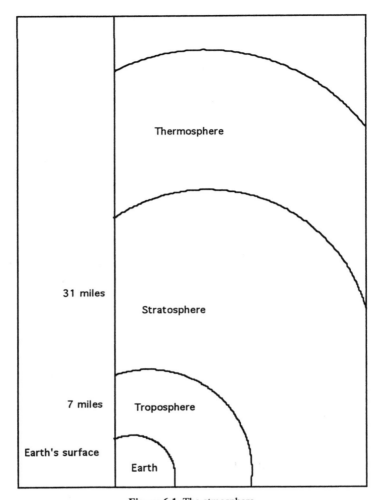

**Figure 6.1** The atmosphere.

## 6.2.1 THE STRATOSPHERE

In the rarefied air of the stratosphere, the significant gas is ozone (life-protecting ozone—not to be confused with pollutant ozone), which is produced by the intense ultraviolet radiation from the sun. In quantity, the total amount of ozone in the atmosphere is so small that, if it were compressed to a liquid layer over the globe at sea level, it would have a thickness of less than 3/16 in.

Ozone contained in the stratosphere can also impact (add to) ozone in the troposphere. Normally, the troposphere contains about 20 parts per billion of ozone. On occasion, however, via the jet stream, this concentration can increase to 5–10 times higher than average.

## 6.3 THE TROPOSPHERE

Extending above earth approximately 27,000 feet, the troposphere is the focus of this text because life on earth (and ozone in the stratosphere) depends on this thin layer of gases. Of more immediate consequence to us is that the formation of all of the Earth's weather takes place within the troposphere.

### 6.3.1 WATER VAPOR

It was pointed out earlier that the gases that are so important to life on Earth are primarily contained in the troposphere. Also note that another important substance is contained in the troposphere: water vapor. Along with being the most remarkable of the trace gases contained in the troposphere, water vapor is also the most variable. Considerable attention is paid to water vapor in the next few chapters because water vapor controls the behavior of most of the air surrounding Earth. Unlike the other trace gases in the atmosphere, water vapor alone exists in gas, solid, and liquid forms. It also functions to add and remove heat from the air when it changes from one form to another.

Water vapor (in combination with air-borne particles, obviously) is essential for the stability of Earth's ecosystem. This water vapor–particle combination interacts with the global circulation of the atmosphere and produces the world's weather, including clouds and precipitation.

## 6.4 A JEKYLL-AND-HYDE VIEW OF THE ATMOSPHERE

When non-city dwellers look up into that great natural canopy above their heads, they see many features provided by our world's atmosphere that we know and enjoy: the blueness and clarity of the sky, the color of a rainbow, the spattering of stars reaching every corner of blackness, and the magical colors of a sunset. The air they breathe carries the smell of ocean air, the refreshing breath of clean air after a thunderstorm, and the beauty contained in a snowflake.

But the atmosphere sometimes presents another face—Mr. Hyde's face. The terrible destructiveness of a hurricane, a tornado, monsoon, typhoon, or a hailstorm; the wearying monotony of winds carrying dusts; rampaging windstorms carrying fire up a hillside—these are some of the terrifying aspects of the face of the disturbed atmosphere.

The atmosphere also presents a Hyde-like face whenever man is allowed to put his pollution into it. The view of the sky above us afforded from patches here and there that are not blocked by buildings can be affected by pollution that masks the stars and makes the visible sky a dirty yellow-brown or, at best, a sickly pale blue.

Fortunately, Earth's atmosphere is self-healing. Air cleaning is provided by clouds and the global circulation system, which constantly purge the air of pollutants. Only when air pollutants overload Nature's way of rejuvenating its systems to their natural state are we faced with the repercussions that can be serious, even life threatening.

## 6.5 ATMOSPHERIC PARTICULATE MATTER

Along with gases and water vapor, Earth's atmosphere is literally a boundless arena for particulate matter of many sizes and types. Atmospheric particulates vary in size from 0.0001 to 10,000 microns. Particulate size and shape have direct bearing on visibility. For example, a spherical particle in the 0.6-micron range can effectively scatter light in all directions, reducing visibility.

The types of airborne particulates in the atmosphere vary widely, with the largest sizes derived from volcanoes, tornadoes, waterspouts, burning embers from forest fires, seed parachutes, spider webs, pollen, soil particles, and living microbes.

The smaller particles (the ones that scatter light) include fragments of rock, salt and spray, smoke, and particles from forested areas. The largest portion of airborne particulates are invisible. They are formed by the condensation of vapors, chemical reactions, photochemical effects produced by ultraviolet radiation, and ionizing forces that come from radioactivity, cosmic rays, and thunderstorms.

Airborne particulate matter is produced either by mechanical weathering, breakage and solution, or by the vapor-to-condensation-to-crystallization process (typical of particulates from a furnace of a coal-burning power plant).

As you might guess, anything that goes up must eventually come down. This is typical of airborne particulates also. Fallout of particulate matter depends, obviously, mostly on their size—less obviously on their shape, density, weight, airflow, and injection altitude. The residence time of particulate matter also is dependent on the atmosphere's cleanup mechanisms (formation of clouds and precipitation) that work to remove them from their airborne suspended state.

Some large particulates may only be airborne for a matter of seconds or minutes. Intermediate sizes may be able to stay afloat for hours or days. The finer particulates may stay airborne for a much longer duration: for days, weeks, months, and even years.

Particles play an important role in atmospheric phenomena. For example, particulates provide the nuclei upon which ice particles and cloud condensation are formed, and they are essential for condensation to take place. The most important role airborne particulates play is in cloud formation. Simply put, without clouds, life would be much more difficult, and the cloudbursts that eventually erupted would cause devastation so extreme that it is hard to imagine or contemplate.

## 6.6 SUMMARY

The situation just described could also result whenever massive forest fires and volcanic action take place. These events would release a superabundance of cloud condensation nuclei, which would overseed the clouds, causing massive precipitation to occur.

If natural phenomena such as forest fires and volcanic eruptions can overseed clouds and cause massive precipitation, then what effect would result from man-made pollutants entering the atmosphere at unprecedented levels? This question and other pollution-related questions will be answered in subsequent chapters.

# CHAPTER 7

# Moisture in the Atmosphere

*Hath the rain a father? or who hath begotten the drops of dew? Out of whose womb came the ice? and the hoary frost of heaven, who hath gendered it? . . . Canst thou lift up thy voice to the clouds, that abundance of water may cover thee? (Job 38:28–29, 34)*

## 7.1 INTRODUCTION

O N a hot day when clouds build up, signifying that a storm is imminent, we do not always appreciate or even know what is happening.

This cloud build-up actually signals that one of the most vital processes in the atmosphere is occurring: the condensation of water as it is raised to higher levels and cooled within strong updrafts of air created either by convection currents, turbulence, or physical obstacles such as mountains. The water originated from the surface—evaporated from the seas and the soil or transpired by vegetation. Once in the atmosphere, however, a variety of events combine to convert the water vapor (produced by evaporation) to water droplets. The air must rise and cool to its dew point, of course. At dew point, water condenses around minute airborne particulate matter to make tiny cloud droplets that form clouds—clouds from which precipitation occurs.

Whether created by the sun heating up a hillside, by jet aircraft exhausts, or factory chimneys, there are actually only ten major cloud types. The deliverers of countless millions of tons of moisture from the Earth's atmosphere, clouds form even from the driest desert air containing as little as 0.1% water vapor. They not only provide a visible sign of atmospheric motion but also indicate change in the atmosphere, portending weather conditions that may be expected up to 48 hours ahead. In this chapter we look at the nature and consequences of these cloud-forming processes.

## 7.2  CLOUD FORMATION

The atmosphere is a highly complex system, and the effects of changes in

121

any single property tend to be transmitted to many other properties. The most profound effect on the atmosphere is caused by alternate heating and cooling of the air, which causes adjustments in relative humidity and buoyancy; they cause condensation, evaporation, and cloud formation.

The temperature structure of the atmosphere (along with other forces that propel the moist air upward) is the main force behind the form and size of clouds. Exactly how does temperature affect atmospheric conditions? For one thing, temperature (that is, heating and cooling of the surface atmosphere) causes vertical air movements. Let's take a look at what happens when air is heated.

Let's start with a simple parcel of air in contact with the ground. As the ground is heated, the air in contact with it will warm also. This warm air increases in temperature and expands. Remember, gases expand when heated much more than liquids or solids, so this expansion is quite marked. In addition, as the air expands, its density falls (meaning that the same mass of air now occupies a larger volume). You've heard that warm air rises? Because of its lessened density, this parcel of air is now lighter than the surrounding air and tends to rise. Conversely, if the air cools, the opposite occurs—it contracts, its density increases, and it sinks. Actually, alternate heating and cooling are intimately linked with the process of evaporation, condensation, and precipitation.

But how does a cloud actually form? Let's look at another example.

On a sunny day, some patches of ground warm up more quickly than others because of differences in topography (soil and vegetation, etc.). As the surface temperature increases, heat passes to the overlying air. Later, by mid-morning, a bulbous mass of warm, moisture-laden air rises from the ground. This mass of air cools as it meets lower atmospheric pressure at higher altitudes. If cooled to its dew point temperature, condensation follows, and a small cloud forms. This cloud breaks free from the heated patch of ground and drifts with the wind. If it passes over other rising air masses, it may grow in height. The cloud may encounter a mountain and be forced higher still into the air. Condensation continues as the cloud cools and if the droplets it holds become too heavy, they fall as rain.

### 7.2.1 MAJOR CLOUD TYPES

Earlier, it was mentioned that there are ten major cloud types. These include:

(1) *Stratiform* genera: *Cirrus, Cirrostratus, Cirrocumulus, Altostratus, Altocumulus, Stratus, Stratocumulus,* and *Nimbostratus*

(2) *Cumuliform* genera: *Cumulus* and *Cumulonimbus*

The cloud groups are split up into genera and then into species, similar to plant classification (including Latin names). The stratiform types are subdivided according to the height of their formation. Cumuliform clouds are the result of local convection. Figure 7.1 shows many of the various cloud forms. Let's take a closer look at each of these cloud types.

A *stratus* cloud is a featureless, gray, low-level cloud. Its base may obscure hilltops or occasionally extend right down to the ground, and because of its low altitude, it appears to move very rapidly on breezy days. Stratus clouds can produce drizzle or snow, particularly over hills, and may occur in huge sheets covering several thousand miles.

*Cumulus* clouds also seem to scurry across the sky, reflecting their low altitude. These small dense, white, fluffy, flat-based clouds are typically short-lived, lasting no more than 10–15 minutes before dispersing. They are typically formed on sunny days, when localized convection currents are set up: these currents can form over factories or even brush fires, which may produce their own clouds.

Cumulus clouds may expand into low-lying, horizontally layered, mas-

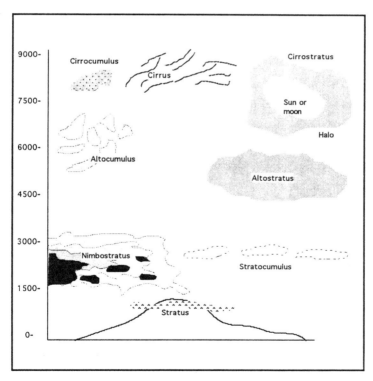

**Figure 7.1** Diagram of stratiform cloud types.

sive *stratocumulus* or into extremely dense, vertically developed, giant *cumulonimbus* with a relatively hazy outline and a glaciated top, sometimes up to seven miles in diameter. These clouds typically form on summer afternoons; their high, flattened tops contain ice, which may fall to the ground in the form of heavy showers of rain or hail.

Rising to middle altitudes, the bluish-gray, layered *altostratus* and rounded, fleecy, whitish-gray *altocumulus* appear to move slowly because of their greater distance from the observer.

*Cirrus* (meaning tuft of hair) clouds are made up of white narrow bands of thin, fleecy parts, are relatively common over northern Europe, and generally ride the jet stream rapidly across the sky.

*Cirrocumulus* clouds are high-altitude clouds composed of a series of small, regularly arranged cloudlets in the form of ripples or grains; they are often present with cirrus clouds in small amounts.

*Cirrostratus* clouds are high-altitude, thin, hazy clouds, usually covering the sky and giving a halo effect surrounding the sun or moon.

## 7.3 MOISTURE IN THE ATMOSPHERE

Let's summarize the information related to how moisture accumulates in and precipitates from the atmosphere. The process of evaporation (converting moisture into vapor) supplies moisture into the lower atmosphere. The prevailing winds then circulate the moisture and mix it with drier air elsewhere.

Water vapor is only the first stage of the precipitation cycle; the vapor must be converted into liquid form. This is usually achieved by cooling, either rapidly, as in convection, or slowly, as in cyclonic storms. Mountains also cause uplift, but the rate will depend upon their height and shape and the direction of the wind.

To actually produce precipitation, the cloud droplets must become large enough to reach the ground without evaporating. The cloud must possess the right physical properties to enable the droplets to grow.

If the cloud lasts long enough for growth to take place, then precipitation will usually occur. Precipitation results from a delicate balance of counteracting forces, some leading to droplet growth and others to droplet destruction.

## 7.4 SUMMARY

Air and water vapor maintain a close kinship; they form a delicately balanced and essential cycle, a relationship essential to Earth's support of life. The mechanisms through which nature carries air and water through their endless, cleansing cycles create (among other things) clouds that supply us with precipitation, which is discussed in the next chapter.

# CHAPTER 8

# Precipitation and Evapotranspiration

*Because it determines the intensity and distribution of many of the processes operating within the system, precipitation is one of the most important regulators of the hydrological cycle. The rate of evapotranspiration is closely related to precipitation and thus is also an integral part of the hydrological cycle.*

## 8.1 INTRODUCTION

**P**RECIPITATION is found in a variety of forms. Which form actually reaches the ground depends upon many factors: for example, atmospheric moisture content, surface temperature, intensity of updrafts, and method and rate of cooling.

Evaporation and transpiration are complex processes that return moisture to the atmosphere. The rate of evapotranspiration depends largely on two factors: (1) how saturated (moist) the ground is and (2) the capacity of the atmosphere to absorb the moisture. In this chapter we discuss the factors responsible for both precipitation and evapotranspiration.

## 8.2 PRECIPITATION

In Chapter 7 it was stated that, if all the essentials were present, precipitation occurs when the dew point is reached. However, Chapter 7 also pointed out that it is quite possible for an air mass or cloud containing water vapor to cool below the dew point without precipitation occurring. In this state, the air mass is said to be *supercooled.*

How, then, are droplets of water formed? Water droplets form around microscopic foreign particles already present in the air. These particles on which the droplets form are called *hygroscopic nuclei.* They are present in the air primarily in the form of dust, salt from seawater evaporation, and from combustion residue. These foreign particles initiate the formation of droplets that eventually fall as precipitation. To have precipitation, larger droplets or drops must form. This may be brought about by two processes: (1) coalescence (collision) or (2) the Bergeron process.

## 8.2.1 COALESCENCE

Simply put, *coalescence* is the fusing together of smaller droplets into larger ones. The variation in the size of the droplets has a direct bearing on the efficiency of this process. Raindrops come in different sizes and can reach diameters up to 7 mm. Having larger droplets greatly enhances the coalescence process.

But what actually goes on inside a cloud to cause rain to fall? To answer this question, we must take a look inside a cloud to see exactly what processes occur to make rain—rain that actually falls as rain. Rainmaking is based on the essentials of the Bergeron process.

## 8.2.2 BERGERON PROCESS

Named after the Swedish meteorologist who suggested it, the *Bergeron process* is probably the most important process for the initiation of precipitation. To gain an understanding of how the Bergeron process works, let's look at what actually goes on inside a cloud to cause rain.

Within a cloud made up entirely of water droplets, there will be a variety of droplet sizes. The air rises within the cloud anywhere from 10 to 20 cm per second (depending on the type of cloud). As the air rises, the drops become larger through collision and coalescence; many will reach drizzle size. Then the updraft intensifies up to 50 cm per second (and more), which reduces the downward movement of the drops, allowing them more time to become even larger. When the cloud becomes approximately 1 km deep, small raindrops of 700 μm diameter are formed.

The droplets, because of their small size, do not freeze immediately, even when the temperatures fall below 0°C. Instead, the droplets remain unfrozen in a supercooled state. However, when the temperature drops as low as −10°C, ice crystals may start to develop among the water droplets. This mixture of water and ice would not be particularly important but for a peculiar property of water. The saturation vapor pressure curve for ice is slightly different from that of water. The air can be saturated for ice when it is not saturated for water. Therefore at −10°C, air saturated with respect to liquid water is supersaturated relative to ice by 10% and at −20°C by 21%. Thus ice crystals in the cloud tend to grow and become heavier at the expense of the water droplets.

Eventually, the ice crystals sink to the lower levels of the cloud where temperatures are only just below freezing. When this occurs, they tend to combine (the supercooled droplets of water act as an adhesive) and form snowflakes. When the snowflakes melt, the resulting water drops may grow further by collision with cloud droplets before they reach the ground as rain. The actual rate at which water vapor is converted to raindrops depends on three main factors: (1) the rate of ice crystal growth, (2) supercooled vapor, and (3) the strength of the updrafts (mixing) in the cloud. The Bergeron rainmaking process is illustrated in Figure 8.1.

### 8.2.3 TYPES OF PRECIPITATION

We stated that in order for precipitation to occur, water vapor must condense, which occurs when water vapor ascends and cools. Three mechanisms by which air rises are convectional, orographic, and frontal. These are illustrated in Figures 8.2–8.4.

#### 8.2.3.1 Convectional Precipitation

*Convectional precipitation* is the spontaneous rising of moist air caused

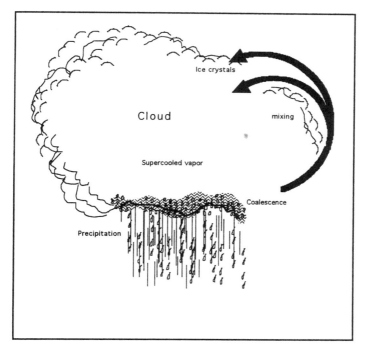

**Figure 8.1** Bergeron rainmaking process. (Adapted from Shipman et al., *An Introduction to Physical Science*, p. 427, 1987.)

by instability (see Figure 8.2). This type of precipitation usually occurs in the summer because localized heating is required to initiate the convection cycle. We have discussed that upward-growing clouds are associated with convection. Since the updrafts are usually strong, cooling of the air is rapid and much water can be condensed quickly, usually confined to a local area, and a sudden summer downpour may occur as a result.

### 8.2.3.2 Orographic Precipitation

*Orographic precipitation* is characteristic of mountainous regions; almost all mountain areas are wetter than the surrounding lowlands. This type of

**Figure 8.2** Convectional precipitation.

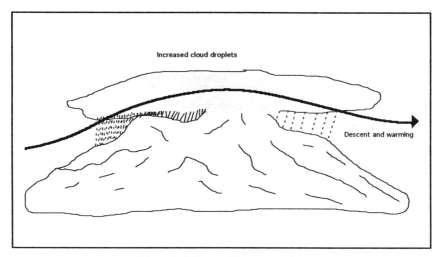

**Figure 8.3** Orographic precipitation.

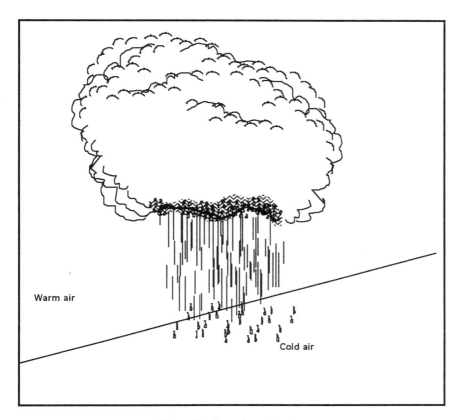

**Figure 8.4** Frontal precipitation.

precipitation occurs when air is forced to rise over a mountain or mountain range. The wind, blowing along the surface of the Earth, ascends along topographic variations. Where air meets this extensive barrier, it is forced to rise. This ascending wind usually gives rise to cooling and encourages condensation, and thus orographic precipitation on the windward side of the mountain range (see Figure 8.3).

### 8.2.3.3 Frontal Precipitation

*Frontal precipitation* results when two different fronts (or the boundary between two air masses characterized by varying degrees of precipitation), at different temperatures, meet. The warm air mass (since it is lighter) moves up and over the colder air mass. The cooling is usually less rapid than in the vertical convection process because the warm air mass moves up at an angle, more of a horizontal motion (see Figure 8.4).

## 8.3 EVAPOTRANSPIRATION

Another important part or process of the hydrological cycle (though it is often neglected because it can rarely be seen) is *evapotranspiration.* More complex than precipitation, evaporation and transpiration is a land–atmosphere interface process whereby a major flow of moisture is transferred from ground level to the atmosphere. It returns moisture to the air, replenishing moisture lost by precipitation, and it also takes part in the global transfer of energy. The rate of evapotranspiration depends largely on two factors: (1) how moist the ground is and (2) the capacity of the atmosphere to absorb the moisture. Therefore the greatest rates are over the tropical oceans, where moisture is always available and the long hours of sunshine and steady trade winds evaporate vast quantities of water.

Just how much moisture is returned to the atmosphere by transpiration? Table 8.1 makes clear, for example, that in the United States, about two-thirds of the average rainfall over the U.S. mainland is returned to the atmosphere via evaporation and transpiration.

### 8.3.1 EVAPORATION

*Evaporation* is the process by which a liquid is converted into a gaseous state. Evaporation takes place (except when air reaches saturation at 100% humidity) almost on a continuous basis. It involves the movement of individual water molecules from the surface of Earth into the atmosphere, a process occurring whenever a vapor pressure gradient exists from the surface to the air (i.e., whenever the humidity of the atmosphere is less than that of

TABLE 8.1. Water Balance in the United States (in bgd*).

| | |
|---|---|
| Precipitation | 4200 |
| Evaporation and transpiration | 3000 |
| Runoff | 1250 |
| Withdrawal | 310 |
| Irrigation | 142 |
| Industry (utility cooling water) | 142 |
| Municipal | 26 |
| Consumed (irrigation loss) | 90 |
| Returned to streams | 220 |

*bgd = billion gallons per day.
*Source:* National Academy of Sciences, 1962.

the ground). Evaporation also requires energy (derived from the sun or from sensible heat from the atmosphere or ground): $2.48 \times 10^6$ Joules to evaporate each kilogram of water at 10°C.

## 8.3.2 TRANSPIRATION

A related process, *transpiration* is the loss of water from plants by evaporation. Most water is lost from the leaves through pores known as stomata, whose primary function is to allow gas exchange between the plant's internal tissues and the atmosphere. Transpiration from the leaf surfaces causes a continuous upward flow of water from the roots via the xylem, which is known as the transpiration stream (see Figure 8.5).

Transpiration occurs mainly by day when the stomata open up under the influence of sunlight. Acting as evaporators, they expose the pure moisture (the plant's equivalent of perspiration) in the leaves to the atmosphere. If the vapor pressure of the air is less than that in the leaf cells, the water is transpired.

As you might guess, because of transpiration, far more water passes through a plant than is needed for growth. In fact, only about 1% or so is actually used in plant growth. Nevertheless, the excess movement of moisture through the plant is important to the plant because the water acts as a solvent, transporting vital nutrients from the soil into the roots and carrying them through cells of the plant. Obviously, without this vital process, plants would die.

## 8.3.3 EVAPOTRANSPIRATION: THE PROCESS

Although evapotranspiration plays a vital role in cycling water over Earth's land masses, it is seldom appreciated. In the first place, distinguishing between evaporation and transpiration is often difficult. Both processes

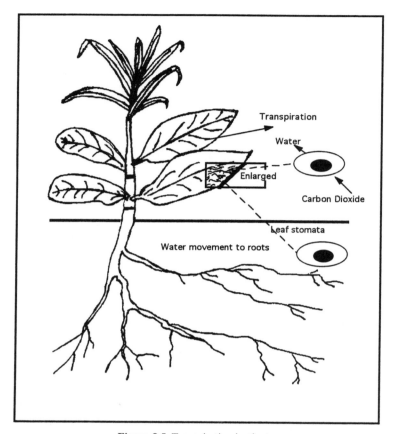

**Figure 8.5** Transpiration in plants.

tend to operate together, so the two are normally combined to give the composite term *evapotranspiration*.

Governed primarily by atmospheric conditions, energy is needed to power the process. Wind also plays an important role, which acts to mix the water molecules with the air and transport them away from the surface. The primary limiting factor in the process is lack of moisture at the surface (soil is dry). Evaporation can continue only so long as a vapor pressure gradient exists between the ground and the air.

## 8.4 SUMMARY

The constant transportation of water in, up, out, around, and through Earth's soil, waterways, and air is an essential part of what keeps our air in motion—generating power, driving weather changes, and perpetuating the essential cycles of our world. We discuss how air keeps in motion in Chapter 9.

## 8.5 REFERENCES

Shipman, J. T., Adams, J. L., & Wilson, J. D., *An Introduction to Physical Science.* Lexington, MA: D.C. Heath and Company, 1987.

Spellman, F. R., *The Science of Water.* Lancaster, PA: Technomic Publishing Co. Inc., 1998.

National Academy of Sciences, *National Research Council Publication 100–B,* 1962.

# The Atmosphere in Motion

*There are scientists and engineers out there in the real world who will tell us that perpetual motion in or for any machine is a pipedream—it's wishful thinking and it is impossible. Have you ever pondered the most dynamic machine of them all—Earth's atmosphere?*

## 9.1 INTRODUCTION

**H**AVE you ever wondered why the Earth's atmosphere is in perpetual motion? It must be in a state of perpetual motion because it constantly strives to eliminate the constant differences in temperature and pressure between different parts of the globe. How are these differences eliminated or compensated for? By its motion, which produces winds and storms. In this chapter the horizontal movements that transfer air around the globe are considered.

## 9.2 GLOBAL AIR MOVEMENT

Basically, winds are the movement of the Earth's atmosphere, which by its weight exerts a pressure on the Earth that we can measure using a barometer. Winds are often confused with air currents, but they are different. Wind is the horizontal movement of air or motion along the Earth's surface. Air currents, on the other hand, are vertical air motions collectively referred to as updrafts and downdrafts.

Throughout history, man has been both fascinated by and frustrated by winds. Man has written about winds almost from the time of the first written word. For example, Herodotus (and later Homer and many others) wrote about winds in his *The Histories.* Wind has had such an impact upon human existence that man has given winds names that describe a particular wind, specific to a particular geographical area. Table 9.1 lists some of these winds, their ancient names, and the regions where they occur. Some of these names

**135**

TABLE 9.1.  Assorted Winds of the World.

| Wind Name | Location |
|-----------|----------|
| aajej | Morocco |
| alm | Yugoslavia |
| biz roz | Afghanistan |
| haboob | Sudan |
| imbat | North Africa |
| datoo | Gibraltar |
| nafhat | Arabia |
| besharbar | Caucasus |
| Samiel | Turkey |
| tsumuji | Japan |
| brickfielder | Australia |
| chinook | America |
| williwaw | Alaska |

are more than just colorful—the winds are actually colored. For example, the *harmattan* blows across the Sahara and is filled with red dust; mariners called this red wind the "sea of darkness."

## 9.3  WHY DO WE HAVE WINDS?

On Earth, we have winds because of the basic laws of motion, which state that, in all dynamic situations (like the production of winds), forces are necessary to produce motion and changes in motion. Note that the gases in the atmosphere are subject to two primary forces: (1) gravity and (2) differences caused by temperature variations.

Our foundational understanding of the principles of air motion, especially as it relates to the influence of gravity, is based to a large degree on the work of Isaac Newton. Newton also formulated the two main laws of motion (which apply to other planets with atmospheres). The first states: *a particle will remain at rest or in uniform motion unless acted upon by another force.* The second law states: *the action of a single force upon a particle causes it to accelerate in the direction of the resultant* (see Figure 9.1).

Because air is a mixture of gases, its behavior is also governed by the gas laws (see Chapter 5). Recall that the pressure of a gas is directly proportional to its temperature. Since pressure equals force divided by area ($P = F/A$), a temperature variation in air generally gives rise to a difference in pressure or force. This coincides with the statement we made earlier that one of the primary forces affecting gases (and motion of gases) in the atmosphere is difference in temperatures.

These various forces are particularly important for movement in the atmosphere because forces are continuously acting on particles of air, causing them to accelerate or decelerate and change their direction.

In the sections that follow, a closer look at the forces acting on air is presented.

### 9.3.1 PRESSURE GRADIENT FORCE

Just as a ball rolls down a hill to a lower elevation, air moves down a pressure hill (along the surface of reasonably constant elevation), or pressure gradient, toward a lower pressure level. To give you a better idea of what a pressure gradient is, let's take a closer look at a parcel of air and see what happens to it.

If we have a small parcel of air some distance from the ground, we know that it will be affected by gravity, which tends to pull or attract all matter toward Earth's center. In addition, the air surrounding the parcel of air exerts pressure. Because this pressure is not the same on all sides of the parcel, it affects the parcel's movement. For example, pressure decreases as we move

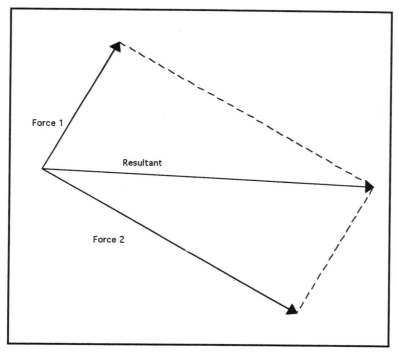

**Figure 9.1** The resultant of two forces acting in different directions. The strength of the force is represented proportionately by the length of the line.

upward in our atmosphere. The force pushing the air upward is greater than the downward force from the overlying atmosphere, which presents a potential upward acceleration of the parcel.

So why doesn't our parcel just fly right on up to the highest level possible? Remember that gravity is also a factor. The vertical upward force is almost exactly balanced by the downward force of gravity—otherwise, we would have lost our atmosphere long ago.

The pressures over a region are mapped by lines drawn through the points (locations) of equal pressure. These points of equal pressure are called *isobars*. There is no air movement along an isobar, because all points on it are of equal pressure. The wind direction is actually at right angles to the isobar in the direction of lower pressure (down the pressure gradient) as shown in Figure 9.2. The magnitude of the force causes the pressure gradient force (movement), and thus the speed of the wind is inversely proportional to the distance between the isobars. It logically follows then that the closer the isobars are together and the more rapidly pressure falls with distance, the stronger is the wind.

You might assume that, since the wind direction is at right angles to the isobar, it flows in that direction. This is not the case, however. The wind actually flows or blows almost parallel to the isobars instead of across them because of the effect of Earth's rotation.

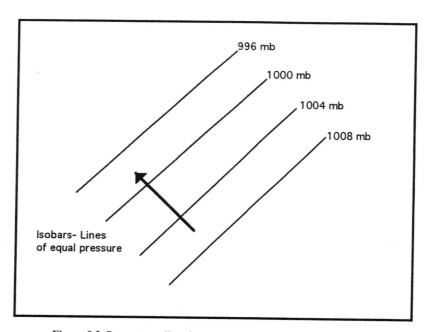

**Figure 9.2** Pressure gradient force acting at right angles to the isobars.

## 9.3.2 CORIOLIS FORCE

Have you ever noticed how water goes down a drain? If so, you may have observed that the water does not flow directly downward but spins clockwise or counterclockwise (depending on whether you are in the northern or southern hemisphere). Have you ever noticed that in the northern hemisphere, rivers scour their banks more severely on the right (viewed along the direction of flow) and that the effect is more evident in high latitudes? Both of these are a consequence of Coriolis force.

*Coriolis force* is the effect of the Earth's rotation on the atmosphere and on all objects on the Earth's surface. In the northern hemisphere, it causes moving objects and currents to be deflected to the right; in the southern hemisphere it causes deflection to the left. The force is named after its discoverer, French mathematician Gaspard Coriolis (1792–1843).

What is the effect of Coriolis force on wind motion? Initially, air moves toward a low-pressure area and away from a high-pressure area. Because of the Coriolis force, the wind is deflected, and in the northern hemisphere, the wind rotates counterclockwise around a low (cyclone) and clockwise around a high (anticyclone). In the southern hemisphere, this action is reversed.

## 9.3.3 FRICTION

Friction also has an impact on wind speed and direction. Flow resistance or drag is caused by internal friction of the wind's molecules or by contact with terrestrial surfaces. The opposing frictional force along a surface is in the opposite direction of the air motion, and its magnitude depends on the air speed.

## 9.4 SUMMARY

Pressure, temperature, and water carry the currents and winds of our world to every corner of the globe. Earth's winds are the indicator of enormously powerful forces. Volcanoes and earthquakes are isolated incidents, but nowhere on the face of the Earth is safe from the forces that wind can bring to bear, or the consequences of that power. That weather and climate, driven by air and water's movement through our atmosphere, affect us all is obvious—and the topic of Chapter 10.

# Weather and Climate

*The Pharisees also with the Sadducees came and tempting desired him that he would show them a sign from heaven. He answered and said unto them, When it is evening ye say, It will be fair weather today for the sky is red and lowering. Oh ye hypocrites, ye can discern the face of the sky, but can ye not discern the signs of the times. (Matthew 16:1–4)*

## 10.1 INTRODUCTION

AN eminent meteorologist once said, "A butterfly flapping its wings in Brazil can cause a tornado in Texas." What the meteorologist was implying is true to a point (and is in line with what some critics might say): because of tiny nuances in Earth's weather patterns, making accurate, long-range weather predictions is extremely difficult.

What is the difference between weather and climate? Some people get these two confused, believing they mean the same thing, but they do not. In this chapter you will gain a clear understanding of the meaning of and difference between the two and also gain a basic understanding of the role weather plays in air pollution.

## 10.2 METEOROLOGY: THE SCIENCE OF WEATHER

*Meteorology* is the science concerned with the atmosphere and its phenomena; the meteorologist observes the atmosphere's temperature, density, winds, clouds, precipitation, and other characteristics and endeavors to account for its observed structure and evaluation (weather, in part) in terms of external influence and the basic laws of physics.

*Weather* is the state of the atmosphere, mainly with respect to its effect upon life and human activities; as distinguished from *climate* (the long-term manifestations of weather), weather consists of the short-term (minutes or months) variations of the atmosphere. Weather is defined primarily in terms of heat, pressure, wind, and moisture.

**141**

At high levels above the Earth, where the atmosphere thins to near vacuum, there is no weather; instead, weather is a near-surface phenomenon. This is evidenced clearly on a day-by-day basis where you see the ever-changing, sometimes dramatic, and often violent weather display.

In the study of air science and, in particular, of air quality, the following determining factors are directly related to the dynamics of the atmosphere, resulting in the local weather. These factors include strength of winds, the direction they are blowing, temperature, available sunlight (needed to trigger photochemical reactions, which produce smog), and the length of time since the last weather event (strong winds or heavy precipitation) cleared the air.

Weather events (such as strong winds and heavy precipitation) that work to clean the air we breathe are beneficial, obviously. However, few people would categorize weather events such as tornadoes, hurricanes, and typhoons as beneficial. Other weather events have both a positive and a negative effect. One such event is El Niño–Southern Oscillation, discussed below.

## 10.2.1 EL NIÑO

El Niño–Southern Oscillation, or ENSO, is a natural phenomenon that occurs every two to nine years on an irregular and unpredictable basis. El Niño is a warming of the surface waters in the tropical eastern Pacific, which causes fish to disperse to cooler waters and, in turn, causes the adult birds to fly off in search of new food sources elsewhere.

Through a complex web of events, El Niño (which means "the child" in Spanish because it usually occurs during the Christmas season off the coasts of Peru and Ecuador) can have a devastating impact on all forms of marine life.

During a normal year, equatorial trade winds pile up warm surface waters in the western Pacific. Thunderheads unleash heat and torrents of rain. This heightens the east–west temperature difference, sustaining the cycle. The jet stream blows from north Asia to California. During an El Niño–Southern Oscillation year, though, trade winds weaken, allowing warm waters to move east. This decreases the east–west temperature difference. The jet stream is pulled farther south than normal, picks up storms it would usually miss, and carries them to Canada or California. Warm waters eventually reach South America.

One of the first signs of its appearance is a shifting of winds along the equator in the Pacific Ocean. The normal easterly winds reverse direction and drag a large mass of warm water eastward toward the South American coastline. The large mass of warm water basically forms a barrier that prevents the upwelling of nutrient-rich cold water from the ocean bottom to the surface. As a result, the growth of microscopic algae that normally flourish in the nutrient-rich upwelling areas diminishes sharply, and that decrease has further repercussions. For example, El Niño–Southern Oscillation has been linked to patterns of subsequent droughts, floods, typhoons, and other costly weather extremes around the globe. Take a look at El Niño–Southern Oscillation's effect on the West Coast of the United States in the winter of 1998, where it has been blamed for West Coast hurricanes, floods, and early snowstorms. On the positive side, ENSO typically brings good news to those who live on the East Coast of the United States: a mild winter and a reduction in the number and severity of hurricanes.

Note that, in addition to reducing the number and severity of hurricanes, in October 1997 the Associated Press reported that a new study has shown that ENSO also deserves credit for invigorating plants and helping to control the pollutants linked to global warming. Researchers have found that El Niño causes a burst of plant growth throughout the world, which removes carbon dioxide from the atmosphere.

Atmospheric carbon dioxide ($CO_2$) has been increasing steadily for decades. The culprits are increased use of fossil fuels and the clearing of tropical rainforests. However, during an ENSO phenomenon, global weather is warmer, there is an increase in new plant growth, and $CO_2$ levels decrease.

Not only does ENSO have a major regional impact in the Pacific, its influence extends to other parts of the world through the interaction of pressure, air flow, and temperature effects.

El Niño–Southern Oscillation is a phenomenon that, although not quite yet completely understood by scientists, causes both positive and negative results, depending upon where you live.

## 10.2.2 THE SUN: THE WEATHER GENERATOR

The sun is the driving force behind weather. Without the distribution and reradiation to space of solar energy, we would experience no weather (as we know it) on Earth. The sun is the source of most of the Earth's heat. Of the gigantic amount of solar energy generated by the sun, only a small portion bombards earth. Most of the sun's solar energy is lost in space. A little over 40% of the sun's radiation reaching earth hits the surface and is changed to heat. The rest stays in the atmosphere or is reflected back into space.

Like a greenhouse, the Earth's atmosphere admits most of the solar radiation. When solar radiation is absorbed by the Earth's surface, it is reradiated as heat waves, most of which are trapped by carbon dioxide and water vapor in the atmosphere, which work to keep the Earth warm in the same way a greenhouse traps heat.

By now you are aware of the many functions performed by the Earth's atmosphere. You should also know that the atmosphere plays an important role in regulating the Earth's heating supply. The atmosphere protects the Earth from too much solar radiation during the day and prevents most of the heat from escaping at night. Without the filtering and insulating properties of the atmosphere, the Earth would experience severe temperatures similar to other planets.

On bright clear nights the Earth cools more rapidly than on cloudy nights because cloud cover reflects a large amount of heat back to Earth where it is reabsorbed.

The Earth's air is heated primarily by contact with the warm earth. When air is warmed, it expands and becomes lighter. Air warmed by contact with Earth rises and is replaced by cold air, which flows in and under it. When this cold air is warmed, it too rises and is replaced by cold air. This cycle continues and generates a circulation of warm and cold air, which is called *convection*.

At the Earth's equator, the air receives much more heat than the air at the poles. This warm air at the equator is replaced by colder air flowing in from north and south. The warm, light air rises and moves poleward high above the Earth. As it cools, it sinks, replacing the cool surface air that has moved toward the equator.

The circulating movement of warm and cold air (convection) and the differences in heating cause local winds and breezes. Different amounts of heat are absorbed by different land and water surfaces. Soil that is dark and freshly plowed absorbs much more heat than grassy fields. Land warms faster than does water during the day and cools faster at night. Consequently, the air above such surfaces is warmed and cooled, resulting in production of local winds.

Winds should not be confused with air currents. Wind is primarily a horizontal air flow. Air currents, on the other hand, are created by air moving upward and downward. Wind and air currents have direct impact on air pollution. Air pollutants are carried and dispersed by wind. An important factor in determining the areas most affected by an air pollution source is wind direction. Since air pollution is a global problem, wind direction on a global scale is important.

Along with wind, another constituent associated with earth's atmosphere is water. Water is always present in the air. It evaporates from the Earth, two-thirds of which is covered by water. In the air, water exists in three states: solid, liquid, and invisible vapor.

The amount of water in the air is called humidity. The *relative humidity* is the ratio of the actual amount of moisture in the air to the amount needed for saturation at the same temperature. Warm air can hold more water than cold. When air with a given amount of water vapor cools, its relative humidity increases; when the air is warmed, its relative humidity decreases.

## 10.2.3 AIR MASSES

An *air mass* is a vast body of air (so large it can have global implications) in which temperature and moisture are much the same at all points in a horizontal direction. An air mass takes on the temperature and moisture characteristics of the surface over which it forms, although its original characteristics tend to persist.

When two different air masses collide, a front is formed. A *cold front* marks the line of advance of a cold air mass from below as it displaces a warm air mass. A *warm front* marks the advance of a warm air mass as it rises up over a cold one.

## 10.3 THERMAL INVERSIONS AND AIR POLLUTION

Earlier, it was pointed out that during the day the sun warms the air near the Earth's surface. Normally, this heated air expands and rises during the day, diluting low-lying pollutants and carrying them higher into the atmosphere. Air from surrounding high-pressure areas then moves down into the low-pressure area created when the hot air rises (see Figure 10.1). This

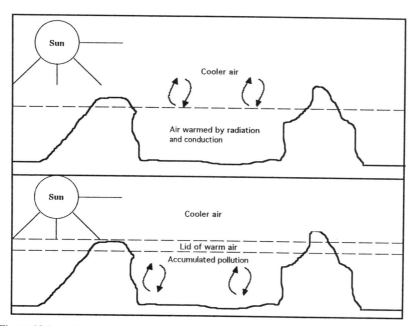

**Figure 10.1** (a) Normal conditions. Air at earth's surface is heated by the sun and rises to mix with the cooler air above it. (b) Thermal inversion. A layer of warm air forms a lid above the earth, and the cooler air at the surface is unable to mix with warm air above. Pollutants are trapped.

continual mixing of the air helps keep pollutants from reaching dangerous levels in the air near the ground.

Sometimes, however, a layer of dense, cool air is trapped beneath a layer of less dense, warm air in a valley or urban basin. This is called a *thermal inversion*. In effect, a warm-air lid covers the region and prevents pollutants from escaping in upward-flowing air currents. Usually these inversions trap air pollutants at ground level for a short period of time. However, sometimes they last for several days when a high-pressure air mass stalls over an area, trapping air pollutants at ground level where they accumulate to dangerous levels.

The best known location in the United States where thermal inversions occur almost on a daily basis is in the Los Angeles basin. The Los Angeles basin is a valley with a warm climate and light winds, surrounded by mountains and located near the Pacific Coast. Los Angeles is a large city with a large population of people and automobiles and possesses the ideal conditions for smog, which is worsened by frequent thermal inversions.

## 10.4  SUMMARY

Mankind has a high degree of self-centeredness, and as the center of our own lives, we occasionally forget that the world as a whole does not revolve around us. When we look at the cold, hard facts of the natural world, we shouldn't be surprised that we can't always have it our own way.

Weather and climate? The only way we can change the conditions around us is to move somewhere else—secure in the knowledge that, even though conditions elsewhere will be different, the new locale will have its own variety of perfect and horrible weather. You love the Tropics? How do you feel about hurricanes?

Climate and weather affect us constantly. We tune in to forecasts to help us choose what we wear, we carry gear to help us through it at the least chance of inclemency, and we alter the temperature in our inside environments to cope with it—a constant lesson in species adaptation to the environment.

# CHAPTER 11

# Microclimates

*Nothing that is can pause or stay;*
*The moon will wax, the moon will wane,*
*The mist and cloud will turn to rain,*
*The rain to mist and cloud again,*
*Tomorrow be today.*
*—Henry Wadsworth Longfellow*

## 11.1 INTRODUCTION

WHEN we think about climate, we are generally referring to generalized weather conditions in a particular region over a period of time. In addition to precipitation and temperature, climates have been classified into zones by vegetation, moisture index, and even by measures of human discomfort. Using the general climate zone names allows us to differentiate between a particular climate (with its specific climatic conditions) in respect to another climate with differing conditions. When geographical patterns in the weather occur again and again over a long period, they can be used to define the climate of a region. Some climate zones are known as the hot climates, which include desert, tropical continental, tropical monsoon, tropical marine, or equatorial types. Warm climates include west coast (Mediterranean) and warm east coast. Another category includes the cool climates such as cold desert, west coast (cool), cool temperature interior, and cool temperate east coast. Finally, there are the mountain and the cold climate categories of cold continental and polar or tundra. Each of these different climate types is differentiated from each other. However, they all have one major feature in common: they are large-scale regional climates (with variations), occurring at various places throughout the world. They only consider the broad similarities of a particular climate at various locations worldwide; local differences are ignored and boundaries are approximate.

What factors determine the variations of climate over the surface of the Earth? The primary factors are (1) the effect of latitude and the tilt of the

Earth's axis to the plane of the orbit about the sun, (2) the large-scale movements of different wind belts over the Earth's surface, (3) the temperature difference between land and sea, (4) the contours of Earth's surface, and (5) the location of the area in relation to ocean currents.

What factors determine how climates are distributed? When considering climate distribution, remember that the world does not fall into compartments. The globe is a mosaic of many different types of climates, just as it is a mosaic of numerous types of ecosystems. The complexity of the distribution of land and sea and the consequent complexity of the general circulation of the atmosphere have a direct effect on the distribution of the climate.

## 11.2 MICROCLIMATES

What is a microclimate? In answering this question, scale must first be talked about. For example, let's take a look at flow of air within a very small environment: the emission of smoke from a chimney. This flow represents one of the smallest spatial subdivisions of atmospheric motion, or microscale weather. On a more realistic, but still relatively small, scale, we must consider the geographical, biological, and man-made features that make local climate different from the general climate. This local climatic pattern is called a *microclimate.*

What are the elements or conditions that cause local climates or microclimates? Location, location, location—and local conditions—are the main ingredients making up a microclimate. Let's look at one example.

Large inland lakes have moderate temperature extremes and climatic differences between the windward and lee sides. For example, Seattle, on the windward side of Lake Washington, and Bellevue, on the lee side only about nine miles east, have microclimatic differences (although modest) between the two cities. These microclimatic differences exist in temperature fluctuations, precipitation levels, wind speed, and in relative humidity.

Even more dramatic differences can be seen in such parameters when a comparison is made between a city such as Milwaukee, on the windward side of Lake Michigan, and Grand Haven, on the lee side, only 85 miles east.

Other examples of microclimates include these listed below:

- near the ground
- over open land areas
- in woodlands or forested areas
- in valley regions
- in hillside regions
- in urban areas
- in seaside locations

In the following sections there is a closer look at these microclimates: at their nature, causative factors, and geographical/topographical locations.

## 11.2.1 MICROCLIMATES NEAR THE GROUND

Nowhere in the atmosphere are climatic differences as distinct as they are near the ground.

For instance, when you go to the beach on a warm summer day, you no doubt have noticed that the grass and water are much cooler to your feet than the sand. So, you may ask, what is it about this area near the ground that produces a microclimate with such major differences?

It's the interface (or activity zone) between the atmosphere and the ground surface (sandy shore) that causes the stark difference in temperature variability. Energy is reaching the sandy beach from the sun and from the atmosphere (though to a much lesser extent). The energy is either reflected and then returned to the atmosphere in a different form or is absorbed and stored in the sandy surface as heat.

Ground level energy absorption is very sensitive to the nature of the ground surface. Ground surface color, wetness, cover (vegetation), and topography are conditions that all affect the interaction between the ground and the atmosphere. Consider snow-covered ground, for example. Clean snow reflects solar radiation, so the surface remains cool and the snow fails to melt. However, dirty snow absorbs more radiation, heats up, and is likely to melt. If the snowy area is shielded by vegetation, the vegetation, too, may protect the snow from the heat of the sun.

A surface cover such as clean snow has the ability to reflect solar radiation because of its high albedo. *Albedo* (the ratio between the light reflected from a surface and the total light falling on it) always has a value less than or equal to 1. An object with a high albedo, near 1, is very bright, while a body with a low albedo, near 0, is dark. For example, freshly fallen snow typically has an albedo that is between 75% and 90+%; that is, 75% to 95% of the solar radiation that is incident on snow is reflected. At the other extreme, the albedo of a rough, dark surface, such as a green forest, may be as low as 5%. The albedos of some common surfaces are listed in Table 11.1. The portion of insolation not reflected is absorbed by the Earth's surface, warming it. This means Earth's albedo plays an important part in the Earth's radiation balance and influences the mean annual temperature and the climate on both local and global scales.

## 11.2.2 MICROCLIMATES OVER OPEN LAND AREAS

Many different properties of ground layer or soil type influence conditions in the thin layer of atmosphere just above it. Light-colored soils do not absorb

TABLE 11.1. The Albedo of Some Surface Types in Percent Reflected.

| Surface | Albedo |
|---|---|
| Water (low sun) | 10–100 |
| Water (high sun) | 3–10 |
| Grass | 16–26 |
| Glacier ice | 20–40 |
| Deciduous forest | 15–20 |
| Coniferous forest | 5–15 |
| Old snow | 40–70 |
| Fresh snow | 75–95 |
| Sea ice | 30–40 |
| Blacktop | 5–10 |
| Desert | 25–30 |
| Crops | 15–25 |

energy as efficiently as do organically rich darker soils. Another important factor is soil moisture. Wet soils are normally dark, but moist soil (because water has a large heat capacity) requires a great deal of energy to raise its temperature. In Figure 11.1 we can clearly see that moist soil warms up more slowly than dry soil.

Soil is a heterogeneous mixture of various particles. In between the soil particles is a large amount of air—air that is a poor conductor of heat. The larger the amount of air between the soil particles, the slower is the heat transfer through the soil. As demonstrated in our example of the sandy beach, on a hot sunny day the heat is trapped in the upper layers, so the surface layers warm up more rapidly and become extremely hot. Water conducts

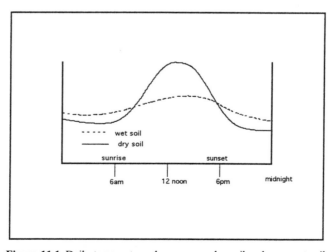

**Figure 11.1** Daily temperature changes over dry soil and over wet soil.

heat more readily than air, so soils that contain some moisture are able to transmit warmth away from the surface more easily than dry soils. This is not always the case, however. If the soil contains too much water, the large heat capacity of the water will prevent the soil from warming, despite heat being conducted from the surface.

## 11.2.3 MICROCLIMATES IN WOODLANDS OR FORESTED AREAS

When you make microclimate comparisons between open land areas and forested areas (commonly referred to as a forest climate), the differences are quite apparent. Forested areas, for example, are generally warmer in winter than open areas, while open land area is warmer in summer than forested areas. The forest climate has reduced wind speeds, while the open land area has higher wind speeds. The forest climate has higher relative humidity, while the open area has lower relative humidity. In the forest climate, water storage capacity is higher and evaporation rates are lower, while in the open land area water storage capacity is lower and evaporation rates are high.

## 11.2.4 MICROCLIMATES IN VALLEY AND HILLSIDE REGIONS

Heavy, cold air flows downhill, forming cold pockets in valleys. Frost is much more common there, so orchards of apples and oranges and vines of grapes are planted on hillsides to ensure frost drainage when cold spells come.

Probably the best way in which to describe the microclimate in a typical valley region is to compare and contrast it with a hillside environment.

In a typical valley region, the daily minimum temperature is much lower than that in a hillside area. The daily and annual temperature range for a valley is much broader than that of a hillside area. In a valley region, more frost occurs than in a hillside region. Windspeed at night is lower in a valley than on a hillside, and morning fog is more prevalent and lasts longer in a valley region.

## 11.2.5 MICROCLIMATES IN URBAN AREAS

The microclimate differences in urban areas compared to those of the countryside are usually quite obvious. A city, for example, is usually characterized by having haze and smog, higher temperatures, lower wind speed, and reduced radiation. The countryside, on the other hand, is characterized by clean, clear air, lower temperatures, and higher wind speeds and radiation.

These different microclimatic conditions should come as no surprise to anyone, especially when you consider what happens when a city is built.

Instead of a mixture of soil or vegetation, the surface layer is covered with concrete, brick, glass, and stone surfaces ranging to heights of several hundred feet. These materials have vastly different physical properties than soil and trees. They shed and carry away water, absorb heat, block and channel the passage of winds, and present albedo levels significantly different from those of the natural world. All of these factors (and more) work to alter the climate conditions in the area.

## 11.2.6 MICROCLIMATES IN SEASIDE LOCATIONS

The major climatic feature associated with seaside locations is the sea breeze. Sea breezes are formed by the different responses to heating of water and land. For example, if we have a bright, sunny morning with little wind, the ground surface warms rapidly as it absorbs short-wave radiation. Most of this heat is retained at the surface, although some will be transferred through the soil. As a result, the temperature of the ground surface increases, and some of the heat warms the air above. When the sun sets, the surface starts to cool rapidly, because there is little storage of heat in the soil. Thus, we find that land surfaces are characterized by high day (and summer) temperatures and low night (and winter) temperatures.

Now let's take a look at the response of the sea, which is very different. Solar energy (sunshine) is able to penetrate through the water to a certain level. Much solar energy has to be absorbed to raise its temperature. Through wave action and convection, the warm surface water is mixed with cooler, deeper water. With enough solar energy and time, the top several feet of water form an active layer where temperature change is slow. Slight warming occurs during the day and slight cooling at night. This means that the sea is normally cooler than the land by day and warmer by night.

The higher temperature over the land by day generates a weak low-pressure area. As this intensifies during daytime heating, a flow of cool, more humid air spreads inland from the sea, gradually changing in strength and direction during the day. At night the reverse occurs, with circulation of air from the cooler land to the warmer sea, though the temperature difference is usually less, so the land breeze is weak. Even large lakes can show a breeze system of this nature.

## 11.3 SUMMARY

Though comparing the local temperature to a place you'd rather be is occasionally of interest, most of us tune into the local weather forecast because we want the information that directly affects us as we try to determine the course of our day.

Microclimate changes become apparent quickly. For example, the country only a few miles away to the south of us may have no snow at all but has freezing rain; to the north they have four more inches of snow than we do. There may be just some drizzle to the west of us—it hasn't really hit there yet—but to the east the ocean is driving in the storm on high winds. Topography, latitude, altitude, natural barricades, the lay of the land relative to prevailing winds, and exposure to the sun all play a part in creating a patchwork quilt of climate conditions that make up the larger—and interconnected—whole.

# CHAPTER 12

# The Endangered Atmosphere: Climate Change

*Humanity is conducting an unintended, uncontrolled, globally perva-
sive experiment whose ultimate consequences could be second only to
nuclear war. The Earth's atmosphere is being changed at an unprecedented rate by pollutants resulting from human activities, inefficient
and wasteful fossil fuel use and the effects of rapid population growth
in many regions. These changes are already having harmful consequences over many parts of the globe. (Toronto Conference statement,
June 1988)*

## 12.1 INTRODUCTION

A RE we headed for warmer times or colder times? Is the greenhouse effect
really happening, and if so, do we really need to worry about it? Are the
tides rising? Does the ozone hole portend disaster in the near future?  Is
Earth's climate changing?

These days, many people are beginning to ask a variety of questions
related to climate change. Such questions seem reasonable when you consider the constant barrage of newspaper headlines, magazine articles and
television news reports we have been exposed to in recent years. Recently,
for example, El Niño (and its devastation of the West Coast of the U.S. and
Peru and Ecuador) has received a lot of media and scientific attention. On
the other side of the coin, it has helped reduce the usual number, magnitude,
and devastation of hurricanes that annually blast the East Coast of the
U.S.—though the people affected by the ice storms in upstate New York and
in Canada and the tornado victims in Florida in the winter of 1998 probably
didn't consider themselves very lucky.

What's going on? What does all this mean? We have plenty of theories
and doomsayers out there, but are they correct? Does anyone really know
the answers? Is there anything we can do?

Yes, we can study the facts, the issues, and the possible consequences, but

**155**

we need to let scientific fact, common sense, and cool-headedness prevail. The main question is, "Will we take the correct action before it is too late?" The key words here are "correct action."

In this chapter, global climate change related to our atmosphere and its problems, actual and potential, are discussed. Consider this: any damage we do to our atmosphere affects the other three mediums—water, soil, and biota. Thus, the endangered atmosphere (if it is endangered) is a major concern to all of us.

## 12.2 THE PAST

What time frames are we referring to when we say "the past"? Table 12.1 gives the entire expanse of time from Earth's beginning to present. Table 12.2 provides the sequence of geological epochs over the past 65 million years, as dated by modern methods. The Paleocene through Pliocene make up the Tertiary Period; the Pleistocene and the Holocene compose the Quaternary Period.

When most people think about climatic conditions in the ancient past, they generally think of two eras: the Ice Age and the period of the dinosaurs. Of course, those two ages take up only a tiny fraction of the time Earth has been spinning around the sun. Let's look at what we know about the past. In the first place, geological history has shown periods when the normal climate of the Earth was so warm that subtropical weather reached to 60° north and south latitude, and there was a total absence of polar ice.

TABLE 12.1. Geological Eras and Periods.

| Era | Period | Millions of Years before Present |
|---|---|---|
| Cenozoic | Quaternary | 2.5–present |
| | Tertiary | 65–2.5 |
| Mesozoic | Cretaceous | 135–65 |
| | Jurassic | 190–135 |
| | Triassic | 225–190 |
| Paleozoic | Permian | 280–225 |
| | Pennsylvanian | 320–280 |
| | Mississippian | 345–320 |
| | Devonian | 400–345 |
| | Silurian | 440–400 |
| | Ordovician | 500–440 |
| | Cambrian | 570–500 |
| | Precambrian | 4,600–570 |

TABLE 12.2. Geological Epochs.

| Epochs | Million Years Ago |
|--------|-------------------|
| Holocene | .01–0 |
| Pleistocene | 1.6–.01 |
| Pliocene | 5–1.6 |
| Miocene | 24–5 |
| Oligocene | 35–24 |
| Eocene | 58–35 |
| Paleocene | 65–58 |

Most people are unaware that, during less than about 1% of Earth's history, glaciers have advanced and reached as far south as what is now the temperate zone of the northern hemisphere. The latest such advance (which started about 1,000,000 years ago) was marked by geological upheaval and (arguably) the beginning of man. During this time, vast ice sheets advanced and retreated over the continents. The following takes a closer look at these Ice Ages.

## 12.2.1  A TIME OF ICE

The oldest known glacial epoch occurred nearly 2 billion years ago. In southern Canada, extending east to west about 1,000 miles, is a series of deposits of glacial origin. Within the last billion years or so, the Earth has experienced at least six major phases of massive, significant climatic cooling and subsequent glaciation, which apparently occurred at intervals of about 150 million years. Each may have lasted as long as 50 million years.

In more recent times (the Pleistocene Epoch to the present), examination of land and oceanic sediment core samples clearly indicates that numerous alterations between warmer and colder conditions have occurred over the last 2 million years (during the middle and early Pleistocene Epoch). At least eight such cycles have occurred in the last million years, with the warm part of the cycle lasting only a relatively short time.

During the Pleistocene Epoch (what we commonly call the Great Ice Age), a series of ice advances began, at times covering over one-fourth of the Earth's land surface with great sheets of ice thousands of feet thick. Glaciers moved across North America many times, reaching as far south as the Great Lakes, and an ice sheet thousands of feet thick spread over Northern Europe, sculpting the land and leaving behind its giant footprints in the form of numerous lakes and swamps and its toeprints in the form of terminal moraines as far south as Switzerland. Evidence appears to indicate that each succeeding glacial advance was more severe than the previous one. The most severe began about 50,000 years ago and ended about 10,000 years ago.

Several glacial advances were separated by interglacial stages, during which the ice melted and temperatures were, on average, higher than today.

"Temperatures were higher than today?" Yes. Keep this important point in mind as we proceed.

Although scientists consider the Earth still to be in a glacial stage (because one-tenth of the globe's surface is still covered by glacial ice), ever since the climax of the last glacial advance, the ice sheet has been retreating, and world climates, although fluctuating, are slowly warming.

How do we know the ice sheet is in a retreating stage? We know from our observations and from well-kept records that clearly show that the last hundred years have seen marked worldwide retreat of ice. Swiss resorts built during the early 1900s to offer scenic views of glaciers now have no ice in sight. Glacier National Park in Montana (world famous for its 50 glaciers and 200 lakes) is not quite the same place it was a hundred years ago. In 1937, a ten-foot pole was driven into the ground at the terminal edge of one of the main glaciers. Today this sign is still in place, but the glacier has retreated several hundred feet back up the slope of the mountain. If this glacial retreat continues and all the ice melts, sea levels would rise more than 200 feet, flooding many of the world's major cities. New York and Boston would become aquariums.

What causes an Ice Age? The cause of these periodic ice ages is a deep enigma in Earth's history. Scientists have advanced many theories ranging from changing ocean currents to sunspot cycles. One fact that is absolutely certain, however, is that an Ice Age event occurs because of a change in Earth's climate. But what brings about such a drastic change? To answer this question, we need to take a closer look at some factors related to climate and climate change.

Climate results from the uneven distribution of heating over the surface of the Earth caused by the Earth's tilt. This tilt is the angle between the Earth's rotational axis and its orbital plane around the sun. Currently, this angle is $23.5°$, but the angle has changed over time (we will discuss this in greater detail shortly).

Long-term climate is also affected by the heat balance of the Earth, which is driven mostly by the concentration of carbon dioxide ($CO_2$) in the atmosphere. Climate change can result if the pattern of solar radiation is changed or if the amount of $CO_2$ changes. Abundant evidence that the Earth does undergo climatic change exists.

Soil core samples and topographical formations show us that the Earth undergoes climatic change. Climate change includes events such as periodic Ice Ages, characterized by glacial and interglacial periods. Major glacial periods lasted up to 100,000 years, with a temperature decrease of about $9°F$, and most of the planet was covered with ice. Minor periods lasted up to 12,000 years with about a $5°F$ decrease in temperature, and ice covered 40°

latitude and above. Smaller periods such as the "Little Ice Age" occurred from about 1000–1850 A.D., when there was about a 3.8°F drop in temperature. Despite its name, "Little Ice Age" was not a true glacial period but, rather, a time of severe winters and violent storms.

We are presently in an interglacial stage that may be reaching its apogee. The Earth has gone through a series of glacial periods, and from the best information available to us, these periods are cyclical. What does that mean to us in the long run? No one knows for sure, but let's look at the effects of Ice Ages.

Ice Ages bring about changes in sea levels. A full-blown Ice Age can change sea level by about 100 meters, which would expose the continental shelves. The exposed continental shelves' composition would be changed because of increased deposition during melt. The hydrological cycle would change because less evaporation would occur. Significant changes in landscape would occur, such as the creation of huge formations on the scale of the Great Lakes. Drainage patterns throughout most of the world would change, with possible massive flooding episodes. Topsoil characteristics would change—glaciers deposit rock and grind away soil.

Are these changes significant? Many areas would be devastatingly affected by the changes that an Ice Age could bring, in particular, Northern Europe, Canada, Seattle, Washington, around the Great Lakes, and near coastal regions.

Let's go back to the main question: What causes Ice Ages? The answer: we are not sure, but we do have some theories. Scientists point out, for example, that to generate a full-blown Ice Age (a massive ice sheet covering most of the globe), certain periodic or cyclic events or happenings would have to take place. The periodic fluctuations referred to would have to affect the solar cycle, for instance. However, we have no definitive evidence that this has ever occurred.

Another theory speculates that periods of widespread volcanic activity generating masses of volcanic dust could block or filter heat from the sun, thus cooling down the Earth.

Some speculate that the carbon dioxide cycle would have to be periodic/cyclic to bring about periods of climate change. References to a so-called factor 2 reduction, causing a 7°F temperature drop worldwide have been made.

Others speculate that another global Ice Age could be brought about by increased precipitation at the poles, caused by changing orientation of continental land masses.

Others theorize that a global Ice Age would result if changes in mean temperatures of ocean currents occurred. But the question is how? By what mechanism?

So what are the most probable causes of Ice Ages on Earth? According to the *Milankovitch hypothesis*, the occurrence of an Ice Age is governed by a combination of factors: (1) the Earth's change of altitude in relation to the sun—the way it tilts in a 41,000-year cycle and at the same time wobbles on its axis in a 22,000-year cycle, making the time of its closest approach to the sun come at different seasons; and (2) the 92,000-year cycle of eccentricity in its orbit round the sun, changing it from an elliptical to a nearly circular orbit, with the severest period of an Ice Age coinciding with the approach to circularity.

So what does all this mean? What this means is we have a lot of speculation about Ice Ages and their causes and their effects. We do not have any speculation about the fact that they actually occurred and that they caused certain things to occur (formation of the Great Lakes, etc.), but there is a lot we do not know—the old "we don't know what we don't know" paradox.

Many possibilities exist. At this point, no single theory is sound, and, doubtless, many factors are involved. But we should keep in mind that we are possibly still in the Pleistocene Ice Age. It may reach another maximum in another 60,000+ years or so. This issue will be revisited again in the next section.

## 12.2.2 WARM WINTER

Maybe you've seen the headlines: "1997 Was the Warmest Year on Record," "Scientists Discover Ozone Hole Is Larger Than Ever," "Record Quantities of Carbon Dioxide Detected in Atmosphere," or "January 1998 Was the Third Warmest January on Record." Have you seen any reports that state that research indicates we are undergoing a cooling trend?

In the previous section, we discussed several possible causes of glaciation and subsequent climatic cooling. In most scenarios discussed, we were left with the old paradox: We don't know what we don't know. Now it's time to discuss how we know what we think we know about climatic change.

### 12.2.2.1 What We Think We Know about Global Climate Change

Two large-scale environmentally significant events took place in 1997: the return of El Niño and the Kyoto Conference: Summit on Global Warming and Climate Change. As to El Niño, 1997 and 1998 news reports have blamed this phenomenon for just about everything and anything that has to do with weather conditions throughout the world. Some of these occurrences are indeed El Niño-related or -generated: the out-of-control fires, droughts, floods, stretches of dead coral, the lack of fish in the water, and few birds around certain Pacific atolls. Few would argue that the devastating storms that struck the west coasts of South America, Mexico, and the U.S. were not El Niño-related. Additionally, few argue against El Niño's effect on the 1997 hurricane season, one of the mildest on record. However, other anomalies or occurrences (such as lower or higher plant growth in certain regions of the globe and absurdities like the appearance of a double rainbow in certain areas) blamed on El Niño certainly are suspect, if not totally ridiculous.

On December 7, 1997, the Associated Press reported that, while delegates at the global climate conference in Kyoto haggled over greenhouse gases and emission limits, a compelling—and so far unanswered—question has emerged: "Is global warming fueling El Niño?"

Nobody is really sure because we need more information, more data than we have today. One thing seems certain, however (based on our paltry amount of recorded data): El Niño is getting stronger and more frequent.

Some scientists fear that the increasing frequency and intensity of El Niño (based on records showing that two of this century's three worst El Niños have come recently, in 1982 and 1997) may be linked to global warming. Experts at the Kyoto Conference say the hotter atmosphere is heating up the world's oceans, which could set the stage for more frequent and extreme El Niños.

We have little doubt that weather-related phenomena seem to be intensifying throughout the globe. Can we be sure that this is related to global warming yet? No. The jury is still out. We need more data, more science, more time.

Is there cause for concern? Yes, there is. According to the Associated Press coverage of the Kyoto Conference, scientist Richard Fairbanks reported that he found some startling evidence of our need for concern. During two months of scientific experiments on Christmas Island (the world's largest atoll in the Pacific Ocean) conducted in the autumn of 1997, he discovered a disturbing scene. The water surrounding the atoll was 7°F higher than average for that time of year, throwing the environmental system out of balance. According to Fairbanks, 40% of the coral was dead, the warmer water had killed off or driven away fish, and the atoll's normally plentiful bird population was almost completely gone.

Few would argue the impact that El Niño is having on the globe. However, we are not certain that it is caused or intensified because of global warming.

The natural question now shifts to, "What do we know about global warming and climate change?"

*USA Today* (December 1997) reported on the results of a report issued by the Intergovernmental Panel on Climate Change and an interview with Jerry Mahlman of the National Oceanic and Atmospheric Administration and Princeton University from which the following information about what most scientists agree on was obtained (p. A-2):

- There is a natural "greenhouse effect." Scientists know how it works, and without it, Earth would freeze.

- The Earth undergoes normal cycles or warming and cooling on grand scales. Ice ages occur every 20,000 to 100,000 years.
- Globally, average temperatures have risen 1°F in the past 100 years, within the range that might occur normally.
- The level of man-made carbon dioxide in the atmosphere has risen 30% since the beginning of the Industrial Revolution in the 19th century and is still rising.
- Levels of man-made carbon dioxide will double in the atmosphere over the next 100 years and generate a rise in global average temperatures of about 3.5°F (larger than the natural swings in temperature that have occurred over the past 10,000 years).
- By 2050, temperatures will rise much higher in northern latitudes than the increase in global average temperatures. Substantial amounts of northern sea ice will melt, and snow and rain in the northern hemisphere will increase. As the climate warms, the rate of evaporation will rise, further increasing warming. Water vapor also reflects heat back to Earth.

### 12.2.2.2  What We Think We Know about Global Warming

What is global warming? To answer this question we need to discuss the greenhouse effect. We know that water vapor, carbon dioxide, and other atmospheric gases (greenhouse gases) help to warm the Earth. Without this greenhouse effect, the Earth's average temperature would be closer to zero than its actual 60°F. As gases are added to the atmosphere, the average temperature could increase, changing orbital climate.

### 12.2.2.2.1  Greenhouse Effect

To understand Earth's Greenhouse Effect, here's an explanation most people (especially gardeners) are familiar with. In a garden greenhouse, the glass walls and ceilings are largely transparent to short-wave radiation from the sun, which is absorbed by the surfaces and objects inside the greenhouse. Once absorbed, the radiation is transformed into long-wave (infrared) radiation (heat), which is radiated back from the interior of the greenhouse. But the glass does not allow the long-wave radiation to escape; instead, it absorbs the warm rays. With the heat trapped inside, the interior of the greenhouse becomes and remains much warmer than the air outside.

The Earth's atmosphere allows much the same greenhouse effect to take place. The short-wave and visible radiation that reaches earth is absorbed by the surface as heat. The long heat waves are then radiated back out toward space, but the atmosphere instead absorbs many of them. This is a natural and balanced process and, indeed, is essential to life as we know it on Earth. The problem comes when changes in the atmosphere radically change the amount of absorption and, therefore, the amount of heat retained. Scientists in recent decades speculate that this may have been happening as various air pollutants cause the atmosphere to absorb more heat. This phenomenon takes place at the local level with air pollution, causing heat islands in and around urban centers.

As pointed out earlier, the main contributors to this effect are the greenhouse gases: water vapor, carbon dioxide, carbon monoxide, methane, volatile organic compounds (VOCs), nitrogen oxides, chlorofluorocarbons (CFCs), and surface ozone. These gases delay the escape of infrared radiation from the Earth into space, causing a general climatic warming. Note that scientists stress that this is a natural process. Indeed, the Earth would be 33°C cooler than it is presently if the "normal" greenhouse effect did not exist (Hansen et al., 1986).

The problem with Earth's greenhouse effect is that human activities are now rapidly intensifying this natural phenomenon, which may lead to global warming. Debate, confusion, and speculation about this potential consequence is rampant. Scientists are not entirely sure whether the recently perceived worldwide warming trend is because of greenhouse gases or because of some other cause or whether it is simply a wider variation in the normal heating and cooling trends they have been studying. If it continues unchecked, however, the process may lead to significant global warming with profound effects. Human impact on the greenhouse effect is real; it has been measured and detected. The rate at which the greenhouse effect is intensifying is now more than five times what it was during the last century (Hansen & Lebedeff, 1989).

### 12.2.2.2.2 Greenhouse Effect and Global Warming

Those who support the theory of global warming base their assumptions on man's altering of the Earth's normal greenhouse effect, which provides the warmth necessary for life. They blame human activities (burning of fossil fuels, deforestation, and use of certain aerosols and refrigerants) for the increased amounts of greenhouse gases. These gases have increased the amounts of heat trapped in the Earth's atmosphere, gradually increasing the temperature of the whole globe.

Many scientists note that (based on recent or short-term observation) the last decade has been the warmest since temperature recordings began in the late 19th century, and that the more general rise in temperature in the last century has coincided with the Industrial Revolution, with its accompanying increase in the use of fossil fuels. Other evidence supports the global warming theory. For example, in the Arctic and Antarctica, places that are synonymous with ice and snow, we see evidence of receding ice and snow cover.

Taking a long-term view, scientists look at temperature variations over thousands or even millions of years. Having done this, they cannot definitively show that global warming is anything more than a short-term variation in Earth's climate. They base this assumption on historical records that have shown the Earth's temperature does vary widely, growing colder with ice ages and then warming again. On another side of the argument, some people point out that the 1980s saw 9 of the 12 warmest temperatures ever recorded, and the Earth's average surface temperature has risen approximately 0.6°C (1°F) in the last century (USEPA, 1995). At the same time, still others offer as evidence that the same decade also saw three of the coldest years: 1984, 1985, and 1986.

So what is really going on? We are not certain, but let's assume that we are indeed seeing long-term global warming. If this is the case, we must determine what is causing it. But here, we face a problem. Scientists cannot be sure of the greenhouse effect's causes. Global warming may simply be part of a much longer trend of warming since the last Ice Age. Though much has been learned in the past two centuries of science, little is actually known about the causes of the worldwide global cooling and warming that have sent the Earth through a succession of major ice ages and smaller ones. We simply don't have the enormously long-term data to support our theories.

### 12.2.2.2.3 Factors Involved with Global Warming/Cooling

Right now, scientists are able to point to six factors that could be involved in long-term global warming and cooling:

(1) Long-term global warming and cooling could result if changes in the Earth's position relative to the sun occur, with higher temperatures when the two are closer together and lower when further apart.

(2) Long-term global warming and cooling could result if major catastrophes occur (meteor impacts or massive volcanic eruptions) and throw pollutants into the atmosphere, blocking out solar radiation.

(3) Long-term global warming and cooling could result if changes in albedo (reflectivity of Earth's surface) occur. If the Earth's surface were more reflective, for example, the amount of solar radiation radiated back toward space instead of absorbed would increase, lowering temperatures on earth.

(4) Long-term global warming and cooling could result if the amount of radiation emitted by the sun changes.

(5) Long-term global warming and cooling could result if the shape and relationship of the land and oceans change.

(6) Long-term global warming and cooling could result if the composition of the atmosphere changes.

This last possibility (if the composition of the atmosphere changes), of course, relates directly to our present concern: have human activities had a cumulative impact large enough to affect the temperature and climate of Earth? We are not certain right now, but we are concerned and alert to the problem.

## 12.3 SUMMARY

If global warming is occurring, we can expect winters to be longer, summers will be warmer, and sea levels will rise about a foot or so in the next hundred years and will continue to do so for many hundreds of years.

Let's take a closer look at the situation. We have routine global temperature measurements for only about 100 years, and these are not too reliable because of changes in instruments and methods of observation.

The only conclusion to be drawn about our climate is that we do not know whether it is changing drastically. The key word is *drastically*. Geologically, we may be at the end of an Ice Age. Evidence indicates that, during interglacial cycles, there is a period when temperatures increase before they plunge. Are we ascending the peak temperature range? How about human impacts on climate? Have they become so marked that we cannot be certain that the natural cycle of Ice Ages (which has lasted for the last 5 million years or so) will continue? Or could we just be having a breathing spell of a few centuries before the next advance of the glaciers?

No one knows for sure.

## 12.4 REFERENCES

Associated Press, in the *Virginian-Pilot* (Norfolk, VA), "Does warming feed El Niño?" p. A-15, December 7, 1997.

Hansen, J. E., et al., "Climate Sensitivity to Increasing Greenhouse Gases," *Greenhouse Effect and Sea Level Rise: A Challenge for This Generation*, ed., M. C. Barth & J. G. Titus. New York: Van Nostrand Reinhold, 1986.

Hansen, J. E. and Lebedeff, G., "Greenhouse Effect of Chloroflurocarbons and Other Trace Gases," *Journal of Geophysical Research 94*, pp. 16,417–16,421, November 1989.

*USA TODAY*, "Global Warming: Politics and economics further complicate the issue," p. A-1,2, December 1, 1997.

USEPA, *Study of Sea Level Rise*, Washington, D.C.: EPA, 1995.

# PART III

# Air Quality

# CHAPTER 13

# Air Quality Fundamentals

*In recent years emphasis has focused on global climate change and its potential repercussions. No doubt, these are important issues that definitely warrant our attention and our concern, but we may have another concern that dwarfs our concern about greenhouse effect and global warming and so forth. Consider this: if we are unable to breathe the air we have now or in the future because of poor air quality, then what difference does it make if we have ozone holes, hotter summers, warmer winters, melting ice caps, and rising tides?*

## 13.1 INTRODUCTION

IN this chapter concepts are covered that will enable us to better understand the anthropogenic impact of pollution on the atmosphere, which in turn will enable us to better understand the key parameters used to measure air quality. Obviously, having a full understanding of air quality is essential. To set the stage for the information to follow in subsequent chapters related to air pollution and air pollution control, we must review a few basic concepts.

## 13.2 EARTH'S HEAT BALANCE

Approximately 50% of the solar radiation entering the atmosphere reaches Earth's surface, either directly or after being scattered by clouds, particulate matter, or atmospheric gases. The other 50% is either reflected directly back or absorbed in the atmosphere, and its energy is reradiated back into space at a later time as infrared radiation. Most of the solar energy reaching the surface is absorbed and must be returned to space to maintain *heat balance*. The energy produced within the Earth's interior (from hot mantle areas via convection and conduction) that reaches the Earth's surface (about 1% of that received from the sun) must also be lost.

Reradiation of energy from the Earth is accomplished by three energy transport mechanisms: radiation, conduction, and convection. *Radiation* of energy, as stated earlier, occurs through electromagnetic radiation in the infrared region of the spectrum. The crucial importance of the radiation mechanism is that it carries energy away from Earth on a much longer wavelength than the solar energy (sunlight) that brings energy to the Earth and, in turn, works to maintain the Earth's heat balance. The Earth's heat balance is of particular interest to us in this text because it is susceptible to upset by human activities.

A comparatively smaller, but significant, amount of heat energy is transferred to the atmosphere by conduction from the Earth's surface. *Conduction* of energy occurs through the interaction of adjacent molecules with no visible motion accompanying the transfer of heat; for example, the whole length of a metal rod will become hot when one end is held in a fire. Because air is a poor heat conductor, conduction is restricted to the layer of air in direct contact with the Earth's surface. The heated air is then transferred aloft by *convection,* the movement of whole masses of air, which may be either relatively warm or cold. Convection is the mechanism by which abrupt temperature variations occur when large masses of air move across an area. Air temperature tends to be greater near the surface of the Earth and decreases gradually with altitude. A large amount of the Earth's surface heat is transported to clouds in the atmosphere by conduction and convection before being lost ultimately by radiation, and this redistribution of heat energy plays an important role in weather and climate conditions.

The Earth's average surface temperature is maintained at about 15°C because of the atmospheric greenhouse effect. Greenhouse effect occurs when the gases of the lower atmosphere transmit most of the visible portion of incident sunlight in the same way as does the glass of a garden greenhouse. The warmed Earth emits radiation in the infrared region, which is selectively absorbed by the atmospheric gases whose absorption spectrum is similar to that of glass. This absorbed energy heats the atmosphere and helps maintain the Earth's temperature. Without this greenhouse effect, the surface temperature would average around −18°C. Most of the absorption of infrared energy is performed by water molecules in the atmosphere. In addition to the key role played by water molecules, carbon dioxide, to a lesser extent, is also essential in maintaining the heat balance. Environmentalists and others concerned with environmental issues are concerned that an increase in the carbon dioxide level in the atmosphere could prevent sufficient energy loss, causing damaging increases in the Earth's temperature. This phenomenon, commonly known as anthropogenic greenhouse effect (see Chapter 12), may occur from elevated levels of carbon dioxide levels caused by increased use of fossil fuels and the reduction in carbon dioxide absorption because of destruction of the rainforest and other forest areas.

## 13.3 BASIC AIR QUALITY

The quality of the air we breathe is not normally a conscious concern to us unless we detect something unusual about the air (its odor, its taste, or that it makes breathing difficult) or have been advised by authorities or the news media that there is cause for concern. Air pollutants in the atmosphere cause great concern because of potential adverse effects on our health.

To have good air quality is a plus for any community. Good air quality attracts industry as well as people who are looking for a healthy place to live and raise a family. It is not unusual to see advertisements that push a locality's "clean or fresh air" as being pollution free.

Note that, although most people do seek an environment that has "clean or fresh air" and is pollution-free to live in, this is not always the case for everyone. A good example of this exception is in the Los Angeles Basin. Before Los Angeles became the megacity it is today, local inhabitants named the Basin area the "Valley of the Smokes" because of the campfires and settlements. This early warning about adverse climatic conditions did not stop settlement, and people today still move to the city. Because of the large number of people who decided to make the LA Basin their home, Los Angeles and California have enacted probably the most restrictive air pollution requirements anywhere.

Air quality is impacted by those things we can see readily by eye (smoke, smog, etc.), by those things that can only be seen under the microscope (pollen, microbes, dust, etc.), and by those substances we can't see (ozone, carbon dioxide, sulfur dioxide, etc.). These compounds are heavily regulated, and it seems that with each passing day the USEPA or other regulatory authority imposes some new regulation for a new or old compound. When you watch a local forecast on television these days, it is not unusual to hear reference to the local "air quality index."

## 13.4 SUMMARY

As previously stated, air pollutants in the air we breathe cause great concern because of potential adverse effects on human health. These adverse health effects include acute conditions such as respiratory difficulties and chronic effects such as emphysema and cancer. Although health concerns related to air pollution are usually at the top of any concerned person's list, we must keep in mind that air pollution has adverse impacts on other aspects of our environment that are important to us such as vegetation, materials, and degradation of visibility.

In any discussion of air quality, certain specific areas must be addressed. For example, any discussion about air quality that does not include a discussion of types of air quality management (regulations), air pollutants, air pollution effects on biodiversity (life), air pollution control technology, and indoor air quality is a hollow effort; thus, these topics and others are discussed in detail in the remaining chapters.

# CHAPTER 14

# Air Quality Management

*This we know: All things are connected like the blood that unites us. We did not weave the web of life, we are merely a strand in it. Whatever we do to the web, we do to ourselves.*

*We love this earth as a newborn loves its mother's heartbeat. If we sell you our land, care for it as we have cared for it. Hold in your mind the memory of the land as it is when you receive it.*

*Preserve the land and the air and the rivers for your children's children and love it as we have loved it. (Chief Seattle, mid-1850s)*

## 14.1 INTRODUCTION

WE have found that to preserve the land and the air and the rivers for our children's children and love it as Chief Seattle and his people did, we must properly manage these valuable and crucial natural resources. We have ignored the danger signs for too long, but over the last few decades we have begun the attempt to control and manage our essential resources.

Proper air quality management includes several different areas related to air pollutants and their control. For example, we can mathematically model to predict where pollutants emitted from a source will be dispersed in the atmosphere and eventually fall to the ground and at what concentration. We have found that pollution control equipment can be added to various sources to reduce the amount of pollutants before they are emitted into the air. We have found that certain phenomena such as acid rain, the greenhouse effect, and global warming are all indicators of adverse effects to the air and other environmental mediums, which result from the excessive amount of pollutants being released into the air. We have found that we must concern ourselves not only with ambient air quality in our local outdoor environment, but also with the issue of indoor air quality.

To accomplish air quality management, we have found that managing is one thing—and accomplishing significant change and improvement is another. We need to add regulatory authority, regulations, and regulatory enforcement to the air quality management scheme; strictly voluntary compliance is ineffective.

We cannot maintain a quality air supply without proper management, regulation, and regulatory enforcement. This chapter presents the regulatory framework governing air quality management. It provides an overview of the environmental air quality laws and regulations used to protect human health and the environment from the potential hazards of air pollution. New legislation, reauthorizations of acts, and new National Ambient Air Quality Standards (NAAQS) have created many changes in the way both government and industry manage their business. Fortunately for our environment and for us, they are management tools that are effective—they are working to manage air quality.

## 14.2 CLEAN AIR ACT (CAA)

When you look at a historical overview of air quality regulations, you might be surprised to discover that most air quality regulations are recent. For example, in the United States, the first attempt at regulating air quality came about through passage of the Air Pollution Control Act of 1955 (Public Law 84–159). This act was a step forward but that's about all; it did little more than move us *toward* effective legislation. Revised in 1960 and again in 1962, the act was supplanted by the Clean Air Act (CAA) of 1963 (Public Law 88–206). CAA 1963 encouraged state, local, and regional programs for air pollution control but reserved the right of federal intervention should pollution from one state endanger the health and welfare of citizens residing in another state. In addition, CAA 1963 initiated the development of air quality criteria upon which the air quality and emissions standards of the 1970s were based.

The move toward air pollution control gained momentum in 1970, first by the creation of the Environmental Protection Agency (EPA) and second by passage of the Clean Air Act of 1970 (Public Law 91–604), for which the EPA was given responsibility for implementation. The act was important because it set primary and secondary ambient air quality standards. Primary standards (based on air quality criteria) allowed for an extra margin of safety to protect public health, while secondary standards (also based on air quality criteria) were established to protect public welfare—animals, property, plants, and materials. Further discussion of these standards is found in Chapter 15.

The Clean Air Act of 1977 (Public Law 95–95) further strengthened the existing laws and set the nation's course toward cleaning up our atmosphere.

In 1990, the president signed the Clean Air Act Amendments of 1990. Specifically, the new law

- encourages the use of market-based principles and other innovative approaches such as performance-based standards and emission banking and trading
- promotes the use of clean low-sulfur coal and natural gas, as well as the use of innovative technologies to clean high-sulfur coal through the acid rain program
- reduces enough energy waste and creates enough of a market for clean fuels derived from grain and natural gas to cut dependency on oil imports by 1 million barrels a day
- promotes energy conservation through an acid rain program that gives utilities flexibility to obtain needed emission reductions through programs that encourage customers to conserve energy

Under CAA 1990 several Titles are listed with specific requirements. For example,

- *Title 1*—specifies provisions for attainment and maintenance of National Ambient Air Quality Standards (NAAQS)
- *Title 2*—specifies provisions relating to mobile sources of pollutants
- *Title 3*—covers air toxins
- *Title 4*—covers specifications for acid rain control
- *Title 5*—addresses permits
- *Title 6*—specifies stratospheric ozone and global protection measures
- *Title 7*—discusses provisions relating to enforcement

## 14.2.1 TITLE 1: ATTAINMENT AND MAINTENANCE OF NAAQS

The Clean Air Act of 1977 has certainly brought about significant improvements in U.S. air quality, but the urban air pollution problems of smog (ozone), carbon monoxide (CO), and particulate matter (PM-10) still persist. For example, currently, over 100 million Americans live in cities that do not attain the public health standards for ozone.

A new, balanced strategy for attacking the urban smog problem was needed. The Clean Air Act of 1990 created this new strategy. Under these new amendments, states are given more time to meet the air quality standard (e.g., up to 20 years for ozone in Los Angeles), but they must make steady, impressive progress in reducing emissions. Specifically, it requires the federal government to reduce emissions from (1) cars, buses, and trucks; (2) consumer products such as window-washing compounds and hair spray; and (3) ships and barges during loading and unloading of petroleum products.

In addition, the federal government must develop the technical guidance that states need to control stationary sources. In urban area air pollution problems of smog (ozone), carbon monoxide (CO), and particulate matter (PM-10), the new law clarifies how areas are designated and redefines "attainment." The EPA is also allowed to define the boundaries of "nonattainment" areas (geographical areas whose air quality does not meet federal air quality standards designed to protect public health). CAA 1990 also establishes provisions defining when and how the federal government can impose sanctions on areas of the country that have not met certain conditions.

For ozone specifically, the new law established nonattainment area classifications ranked according to the severity of the area's air pollution problem. These classifications are

- marginal
- moderate
- serious
- severe
- extreme

The EPA assigns each nonattainment area into one of these categories, thus prompting varying requirements the areas must comply with to meet the ozone standard.

Again, nonattainment areas have to implement different control measures, depending upon their classifications. Those closest to meeting the standard, for example, are the marginal areas, which are required to conduct an inventory of their ozone-causing emissions and institute a permit program. Various control measures must be implemented by nonattainment areas with more serious air quality problems; that is, the worse the air quality, the more controls the area will have to implement.

For carbon monoxide and particulate matter, CAA 1990 also establishes similar programs for areas that do not meet the federal health standard. Areas exceeding the standards for these pollutants are divided into "moderate" and "serious" classifications. Areas that exceed the carbon monoxide standard (i.e., the degree to which they exceed it) are required primarily to implement programs introducing oxygenated rules and/or enhanced emission inspection programs. Likewise, areas exceeding the particulate matter standard have to (among other requirements) implement either reasonably available control measures (RACMs) or best available control measures (BACMs).

Title 1 attainment and maintenance of NAAQS requirements have gone a long way toward improving air quality in most locations throughout the United States. However, on November 27, 1996, in an effort to upgrade NAAQS for ozone and particulate matter, the USEPA proposed the National Ambient Air Quality Standards and later put into effect the two new NAAQS for ozone and particulate matter smaller than 2.5-$\mu$m diameter ($PM_{2.5}$). These rules appear at 62 FR 38651 for particulate matter and 62 FR 38855 for ozone (both July 18, 1997). They are the first update in 20 years for ozone (smog) and the first in 10 years for particulate matter (soot).

Table 14.1 lists the National Ambient Air Quality Standards, including the new requirements. Note that NAAQS are important but not enforceable by themselves. The standards set ambient concentration limits for the protection of human health and environment-related values. However, remember that only in a very rare case will any one source of air pollutants be responsible for the concentrations in an entire area.

TABLE 14.1. National Ambient Air Quality Standards.

| Pollutant/ Averaging Time | Primary Standard[a] | Primary Standard[b] | Comments |
|---|---|---|---|
| *Particulate Matter* PM$_{2.5}$, annual | 15 µg/m$^3$ | | Based on 3-year average of annual arithmetic mean PM$_{2.5}$ concentration from single or multiple community-oriented monitors |
| PM$_{2.5}$, 24-hour | 65 µg/m$^3$ | | Based on 3-year average of 98th percentile of 24-hour PM$_{2.5}$ concentrations at each population-oriented monitor within an area |
| PM$_{10}$, annual[c] | 50 µg/m$^3$ | 50 µm/m$^3$ | Attained when expected annual arithmetic mean ≤50 µg/m$^3$ |
| PM$_{10}$, 24-hour[c] | 150 µg/m$^3$ | 150 µg/m$^3$ | Based on 99th percentile of 24-hour PM$_{10}$ concentrations at each monitor within an area |
| *Sulfur Dioxide* SO$_2$, annual | 80 µg/m$^3$ (0.03 ppm) | — | Never to be exceeded |
| SO$_2$, 24-hour | 365 µg/m$^3$ (0.14 ppm) | — | Not to be exceeded more than once per year |
| SO$_2$, 3-hour | 1,300 µg/m$^3$ | — | Not to be exceeded more than once per year |
| *Nitrogen Dioxide* NO$_2$, annual | 100 µg/m$^3$ (0.053 ppm) | 100 µg/m$^3$ (0.053 ppm) | Never to be exceeded |

*(continued)*

TABLE 14.1 (continued). National Ambient Air Quality Standards.

| Pollutant/ Averaging Time | Primary Standard[a] | Primary Standard[b] | Comments |
|---|---|---|---|
| *Ozone*<br>$O_3$, 8-hour | 157 µg/m³<br>(0.08 ppm) | 157 µg/m³<br>(0.08 ppm) | Based on 3-year average of annual 4th-highest daily maximum 8-hour ozone concentrations |
| $O_3$, 1-hour | 235 µg/m³<br>(0.12 ppm) | 235 µg/m³<br>(0.12 ppm) | Standard is attained when the expected number of exceedances ≤1, to be phased out (but present nonattainment areas must show 3 consecutive years of data meeting 1-hour standard before becoming attainment areas) |
| *Carbon Monoxide*<br>CO, 8-hour | 10 mg/m³<br>(9 ppm) | — | Not to be exceeded more than once per year |
| $CO_3$, 1-hour | 40 mg/m³<br>(35 ppm) | — | Not to be exceeded more than once per year |
| *Lead*<br>Pb, calendar quarter | 1.5 µg/m³ | — | Never to be exceeded |

[a]For the protection of human health, with an adequate margin of safety.
[b]For the protection of other values, such as visibility, crops, materials, etc.
[c]Note that, although the previous ozone standard is to be phased out, the $PM_{10}$ standards remain.
*Source:* USEPA, NAAQS, 1997.

## 14.2.2 TITLE 2: MOBILE SOURCES

Cars, trucks, and buses account for almost half of the emissions (even though great strides have been made since the 1960s in reducing the amounts) of the ozone precursors, volatile organic carbons (VOCs) and nitrogen oxides, and up to 90% of the CO emissions in urban areas. A large portion of the emission reductions gained from motor vehicle emission controls has been offset by the rapid growth in the number of vehicles on the highways and the total miles driven.

Because of the unforeseen growth in automobile emissions in urban areas, compounded with the serious air pollution problems in many urban areas, Congress made significant changes to the motor vehicle provisions of the 1977 Clean Air Act. The Clean Air Act of 1990 established even tighter pollution standards for emissions from motor vehicles. These standards were designed to reduce tailpipe emissions of hydrocarbons, nitrogen oxides, and carbon monoxide on a phased-in basis that began with model year 1994. Automobile manufacturers are also required to reduce vehicle emissions resulting from the evaporation of gasoline during refueling.

The latest Clean Air Act (1990 with 1997 amendments for ozone and particulate matter) also require fuel quality to be controlled. New programs were required for cleaner or reformulated gasoline initiated in 1995 for the nine cities with the worst ozone problems. Other cities were given the option to "buy in" to the reformulated gasoline program. In addition, a clean fuel car pilot program was established in California, which requires the phase-in of tighter emission limits for several thousand vehicles in model year 1996 and up to 300,000 by model year 1999. The law allows these standards to be met with any combination of vehicle technology and cleaner fuels. Note that the standards will become even stricter in 2001.

### 14.2.3 TITLE 3: AIR TOXINS

Toxic air pollutants (those that are hazardous to human health or the environment—carcinogens, mutagens, and reproductive toxins) were not specifically covered under the Clean Air Act of 1977. This situation is quite surprising and alarming when you consider that information generated as a result of SARA Title III (Superfund Section 313) indicates that in the U.S. more than 2 billion pounds of toxic air pollutants are emitted annually.

The Clean Air Act of 1990 offered a comprehensive plan for achieving significant reductions in emissions of hazardous air pollutants from major sources. The new law improved the EPA's ability to address this problem effectively and dramatically accelerated progress in controlling major toxic air pollutants.

The 1990 law includes a list of 189 toxic air pollutants whose emissions must be reduced. The EPA was required to publish a list of source categories that emit certain levels of these pollutants. The EPA was also required to issue maximum achievable control technology (MACT) standards for each listed source category, and the law also established a Chemical Safety Board to investigate accidental releases of extremely hazardous chemicals.

### 14.2.4 TITLE 4: ACID DEPOSITION

This section discusses the acid rain problem. Someone knowledgeable or trained in environmental science might wonder whether the rainfall is as clean and pure as it should be. Is this rainfall carrying acids as strong as lemon juice or vinegar in it, capable of harming both living and nonliving things such as trees, lakes, and man-made structures?

Maybe such a concern was unheard of before the Industrial Revolution, but today, the purity of rainfall is a major concern for many people, especially regarding acidity. Most rainfall is slightly acidic because of decomposing organic matter, the movement of the sea, and volcanic eruptions, but the principal factor is atmospheric carbon dioxide, which causes carbonic acid to form. *Acid rain* (pH < 5.6) is produced by the conversion of the primary pollutants sulfur dioxide and nitrogen oxides to sulfuric acid and nitric acid, respectively. These processes are complex and are dependent on the physical dispersion processes and the rates of the chemical conversions.

Contrary to popular belief, acid rain is not a new phenomenon, nor does it result solely from industrial pollution. Natural processes—volcanic eruptions and forest fires, for example—produce and release acid particles into the air. The burning of forest areas to clear land in Brazil, Africa, and other countries also contributes to acid rain. However, the rise in manufacturing that began with the Industrial Revolution dwarfs all other contributions to the problem.

The main culprits are emissions of sulfur dioxide from the burning of fossil fuels such as oil and coal and nitrogen oxide, formed mostly from internal combustion engine emissions, which is readily transformed into nitrogen dioxide. These mix in the atmosphere to form sulfuric acid and nitric acid.

In dealing with atmospheric acid deposition, the Earth's ecosystems are not completely defenseless; they can deal with a certain amount of acid through natural alkaline substances in soil or rocks that buffer and neutralize acid. The American Midwest and southern England are areas with alkaline soil (limestone and sandstone), which provide some natural neutralization. Areas with thin soil and those laid on granite bedrock, however, have little ability to neutralize acid rain.

Scientists continue to study how plants and animals are damaged or killed by acid rain. This complex subject has many variables. We know from various episodes of acid rain that pollution can travel over very long distances. Lakes in Canada and New York are feeling the effects of coal-burning in the Ohio Valley. For this and other reasons, the lakes of the world are where most of the scientific studies have taken place. In lakes, the smaller organisms often die off first, leaving the larger animals to starve to death; sometimes the larger animals (fish) are killed directly. As lake water becomes more acidic, it dissolves heavy metals, leading to concentrations at toxic and often lethal levels. Have you ever wandered up to the local lake shore and observed thousands of fish belly up? Loss of life in lakes also disrupts the system of life on the land and the air around them.

In some parts of the United States, the acidity of rainfall has fallen well below 5.6. In the northeastern U.S., for example, the average pH of rainfall is 4.6, and rainfall with a pH of 4.0 (1000 times more acidic than distilled water) is not unusual.

Despite intensive research into most aspects of acid rain, scientists still have many areas of uncertainty and disagreement. That is why the progressive, forward-thinking countries emphasize the importance of further research into acid rain. And that is why the Clean Air Act of 1990 was strengthened to initiate a permanent reduction in $SO_2$ levels.

One of the interesting features of the 1990 Act is that it allowed utilities to trade allowances within their systems and/or buy or sell allowances to and from other affected sources. Each source must have sufficient allowances to cover its annual emissions. If not, the source is subject to excess emissions fees and a requirement to offset the excess emissions in the following year. The 1990 law also included specific requirements for reducing emissions of nitrogen oxides for certain boilers.

## 14.2.5 TITLE 5: PERMITS

The 1990 law also introduced an operating permit system similar to the National Pollution Discharge Elimination System (NPDES). The permit system has a twofold purpose: (1) to ensure compliance with all applicable requirements of the CAA and (2) to enhance the EPA's ability to enforce the act. Under the act, air pollution sources must develop and implement the program, and the EPA must issue permit program regulations, review each state's proposed program, and oversee the state's effort to implement any approved program. The EPA must also develop and implement a federal permit program when a state fails to adopt and implement its own program.

## 14.2.6 TITLE 6: OZONE AND GLOBAL CLIMATE PROTECTION

We have already discussed the global climate problem (Chapter 12), but let's take a look at the stratospheric ozone problem.

Ozone is formed in the stratosphere by radiation from the sun, and helps to shield life on earth from some of the sun's potentially destructive ultraviolet (UV) radiation.

In the early 1970s, scientists suspected that the ozone layer was being depleted. By the 1980s, it became clear that the ozone shield was indeed thinning in some places and, at times, even has a seasonal hole in it, notably over Antarctica. The exact causes and actual extent of the depletion are not yet fully known, but most scientists believe that various chemicals in the air are responsible.

Most scientists identify the family of chlorine-based compounds, most notably chlorofluorocarbons (CFCs) and chlorinated solvents (carbon tetrachloride and methyl chloroform) as the primary culprits involved in ozone depletion. In 1974, Molina and Rowland hypothesized the CFCs, which contain chlorine, were responsible for ozone depletion. They pointed out that chlorine molecules are highly active and readily and continually break apart the three-atom ozone into the two-atom form of oxygen generally found close to Earth in the lower atmosphere.

The Interdepartmental Committee for Atmospheric Sciences (1975) estimates that a 5% reduction in ozone could result in nearly a 10% increase in cancer. This already frightening scenario was made even more frightening by 1987, when evidence showed that CFCs destroy ozone in the stratosphere above Antarctica every spring. The ozone hole had become larger, with more than half of the total ozone column wiped out, and essentially all ozone disappeared from some regions of the stratosphere (Davis & Cornwell, 1991).

In 1988, Zurer reported that, on a worldwide basis, the ozone layer shrank approximately 2.5% in the preceding decade. This obvious thinning of the ozone layer has increased the chances of skin cancer and cataracts, and it is also implicated in suppression of the human immune system and damage to other animals and plants, especially aquatic life and soybean crops. The urgency of the problem spurred the 1987 signing of the Montreal Protocol by 24 countries, which required signatory countries to reduce their consumption of CFCs by 20% by 1993 and by 50% by 1998, marking a significant achievement in solving a global environmental problem.

The Clean Air Act of 1990 borrowed from EPA requirements already on the books in other regulations and mandated the phasing out of the production of substances that deplete the ozone layer. Under these provisions, the EPA was required to list all regulated substances along with their ozone-depletion potential, atmospheric lifetime, and global warming potentials.

### 14.2.7 TITLE 7: ENFORCEMENT

A broad array of authorizations are contained within the Clean Air Act to make the law more readily enforceable. The EPA was given new authority to issue administrative penalties with fines and field citations (with fines) for smaller infractions. In addition, sources must certify their compliance, and the EPA has authority to issue administrative subpoenas for compliance data.

### 14.3 SUMMARY

In the management of our natural resources, we seek balance—a balance between preserving the beauty and purity of the incredible wealth of our varied and marvelous natural world and fulfilling the needs, wants, goals, and dreams of our burgeoning human population.

Balance is the answer: control and management—separating our essentials from our desires—and recognizing that compromise is not only acceptable, but essential.

## 14.4 REFERENCES

Davis, M. L. & Comwell, D. A., *Introduction to Environmental Engineering.* New York: McGraw-Hill, 1991.

Molina, M. & Rowland, F., "Stratospheric Sink for Chlorophloryl Methane: Chloride Atom Catalyzes Destruction of Ozone," *Nature,* 248:810–812, 1974.

Zurer, P. S., "Studies on Ozone Destruction Expand beyond Antarctic," *Chemical & Engineering News,* pp. 18–25, May 1988.

# CHAPTER 15

# Air Pollution

*Thousands of chemicals are commonly used through the world for industrial, agricultural, and domestic purposes with many new ones being produced yearly. The majority of these chemicals, many of which are toxic . . . eventually enter into the atmosphere and may pose a risk to the well-being of plants, animals, and microorganisms. (Barker & Tingey, 1991, p. 3)*

## 15.1 INTRODUCTION

IN the past, the sight of belching smokestacks was a comforting sight to many people: more smoke equaled more business, which indicated that the economy was healthy. But many of us are now troubled by evidence that indicates that polluted air adversely affects our health. Many toxic gases and fine particles entering the air pose health hazards: cancer, genetic defects, and respiratory disease. Nitrogen and sulfur oxides, ozone, and other air pollutants from fossil fuels are inflicting damage on our forests, crops, soils, lakes, rivers, coastal waters, and buildings. Chlorofluorocarbons (CFCs) and other pollutants entering the atmosphere are depleting the Earth's protective ozone layer, allowing more harmful ultraviolet radiation to reach the Earth's surface. Fossil fuel combustion is increasing the amount of carbon dioxide in the atmosphere, which can have severe long-term environmental impact.

Historically, many felt that the air renewed itself through interaction with vegetation and the oceans in sufficient quantities to make up for the influx into our atmosphere of anthropogenic pollutants. Today, however, this kind of thinking is being challenged by evidence that clearly indicates that increased use of fossil fuels, expanding industrial production, and growing use of motor vehicles are having a detrimental effect on the atmosphere and the environment. In this chapter, we discuss pollutant dispersal, transformation, and deposition mechanisms and examine the types and sources of air pollutants.

## 15.2 ATMOSPHERIC DISPERSION, TRANSFORMATION, AND DEPOSITION

Air pollutants are released from both stationary and mobile sources. Scientists have gathered much information on the sources, quantity, and toxicity levels of these pollutants. The measurement of air pollution is an important scientific skill, and practitioners of this skill are usually well founded in the pertinent related sciences, in modeling aspects applicable to their studies, and in analyses of air pollutants in the ambient atmosphere. However, to get at the very heart of air pollution, the practitioner must also be well versed in how to determine the origin of the pollutants and understand the mechanics of the pollutants' dispersal, transport, and deposition (see Figure 15.1).

Air pollutant practitioners must constantly deal with one basic fact: air pollutants rarely stay at their release location. Instead, wind flow conditions and turbulence, local topographic features, and other physical conditions work to disperse these pollutants. So, along with having a thorough knowledge and understanding of the pollutants in question, the air pollution practitioner has a need for detailed knowledge of the atmospheric processes that govern their subsequent dispersal and fate. Chapter 10 has already

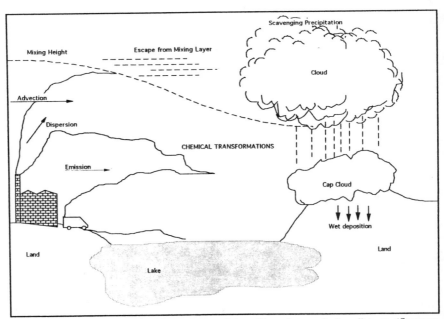

**Figure 15.1** Processes involved in transport and deposition of atmospheric pollutants. (*Source:* Adapted from Barker & Tingey, p. 32, 1991.)

discussed wind and its formation, a critical factor in air pollutant dispersion. In this section, we discuss several other important factors related to pollutant dispersal and fate.

Conversion of precursor substances to secondary pollutants such as ozone is an example of chemical transformation in the atmosphere. Transformations (both physical and chemical) affect the ultimate impact of originally emitted air pollutants.

Pollutants emitted to the atmosphere do not remain there forever. Two common deposition (depletion) mechanisms are *dry deposition* (the removal of both particles and gases as they come into contact with the Earth's surface) and *washout* (the uptake of particles and gases by water droplets and snow and their removal from the atmosphere as precipitation that falls to the ground). Acid deposition (acid rain) is a form of pollution depletion from the atmosphere.

The following sections discuss atmospheric dispersion of air pollutants in greater detail and the main factors associated with this phenomenon, including weather, turbulence, adiabatic lapse rate, mixing, topography, temperature inversions, plume rise, and transport.

## 15.2.1 WEATHER

Recall that in Chapter 10, it was stated that air contained in Earth's atmosphere is not still. Constantly in motion, air masses warmed by solar radiation rise at the equator and spread toward the colder poles where they sink and flow downward, eventually returning to the equator. Near the Earth's surface, as a result of the Earth's rotation, major wind patterns develop. During the day, the land warms more quickly than the sea; at night, the land cools more quickly. Local wind patterns are driven by this differential warming and cooling between the land and adjacent water bodies. Normally, onshore breezes bring cooler, denser air from over the land masses out over the waters during the night. Precipitation is also affected by wind patterns. Warm, moisture-laden air rising from the oceans is carried inland, where the air masses eventually cool, causing the moisture to fall as rain, hail, sleet, or snow.

Even though pollutant emissions may remain relatively constant, air quality varies tremendously from day to day. The determining factors have to do with weather.

Weather conditions have a significant impact on air quality and air pollution, both favorable and unfavorable, especially in local conditions. For example, on hot, sun-filled days, when the weather is calm with stagnating high-pressure cells, air quality suffers because these conditions allow the build-up of pollutants on the ground level. When local weather conditions include cool, windy, stormy weather with turbulent low-pressure cells and cold fronts, conditions allow the upward mixing and dispersal of air pollutants.

Weather has a direct impact on pollution levels in both mechanical and chemical ways. Mechanically, precipitation works to cleanse the air of pollutants (transferring the pollutants to rivers, streams, lakes, or the soil). Winds transport pollutants from one place to another. Winds and storms often dilute pollutants with cleaner air, making pollution levels less annoying in the area of their release. Air and its accompanying pollution (in a low-pressure cell) is also carried aloft by air heated by the sun. When wind accompanies this rising air mass, the pollutants are diluted with fresh air. In a high-pressure cell, the opposite occurs, with air and the pollutants it carries sinking toward the ground. With no wind, these pollutants are trapped and concentrated near the ground where serious air pollution episodes may occur.

Chemically, weather can also affect pollution levels. Winds and turbulence mix pollutants together in a giant chemical broth in the atmosphere. Energy from the sun, moisture in the clouds, and the proximity of highly reactive chemicals may cause chemical reactions, which lead to the formation of secondary pollutants. Many of these secondary pollutants may be more dangerous than the original pollutants.

## 15.2.2 TURBULENCE

In the atmosphere, the degree of turbulence (which results from wind speed and convective conditions related to the change of temperature with height above the Earth's surface) is directly related to stability (a function of vertical distribution of atmospheric temperature). The stability of the atmosphere refers to the susceptibility of rising air parcels to vertical motion; consideration of atmospheric stability or instability is essential in establishing the dispersion rate of pollutants. When specifically discussing the stability of the atmosphere, we are referring to the lower boundary at the Earth's surface, where air pollutants are emitted. The degree of turbulence in the atmosphere is usually classified by stability class. Ambient and adiabatic lapse rates are a measure of atmospheric stability.

Stability is divided into three classes: stable, unstable, and neutral. A *stable atmosphere* is marked by air cooler at the ground than aloft, by low wind speeds, and consequently, by a low degree of turbulence. A plume of pollutants released into a stable lower layer of the atmosphere can remain relatively intact for long distances. Thus, we can say that stable air discourages the dispersion and dilution of pollutants. An *unstable atmosphere* is marked by a high degree of turbulence. A plume of pollutants released into an unstable atmosphere may exhibit a characteristic looping appearance produced by turbulent eddies. A *neutrally stable atmosphere* is an intermediate class between stable and unstable conditions. A plume of pollutants released into a neutral stability condition is often characterized by a coning appearance as the edges of the plume spread out in a V-shape.

The importance of the "state of the atmosphere" and stability's effects cannot be overstated. The ease with which pollutants can disperse vertically into the atmosphere is mainly determined by the rate of change of air temperature with height (altitude). Therefore, air stability is a primary factor in determining where pollutants will travel and how long they will remain aloft. Stable air discourages the dispersion and dilution of pollutants. Conversely, in unstable air conditions, rapid vertical mixing takes place, encouraging pollutant dispersal, which increases air quality.

### 15.2.3 ADIABATIC LAPSE RATE

With an increase in altitude in the troposphere, the temperature of the ambient air usually decreases. The rate of temperature change with height is called the lapse rate. On average, temperature decreases $-0.65°C/100$ m or $-6.5°C/km$. This is the *normal lapse rate.*

In a dry environment, when a parcel of warm, dry air is lifted in the atmosphere, it undergoes adiabatic expansion and cooling. This adiabatic cooling results in a lapse rate of $-1°C/100$ m or $-10°C/km$, the *dry adiabatic lapse rate.*

When the ambient lapse rate exceeds the adiabatic lapse rate, the ambient rate is said to be *superadiabatic,* and the atmosphere is highly unstable. When the two lapse rates are exactly equal, the atmosphere is said to be *neutral.* When the ambient lapse rate is less than the dry adiabatic lapse rate, the ambient lapse rate is termed *subadiabatic,* and the atmosphere is stable.

The cooling process within a rising parcel of air is assumed to be adiabatic (occurring without the addition or loss of heat). A rising parcel of air (under adiabatic conditions) behaves like a rising balloon, with the air in that distinct parcel expanding as it encounters air of lesser density, until its own density, is equal to that of the atmosphere that surrounds it. This process is assumed to occur with no heat exchange between the rising parcel and the ambient air (Peavy et al., 1985).

### 15.2.4 MIXING

Within the atmosphere, for effective pollutant dispersal to occur, turbulent mixing is important. Turbulent mixing, the result of the movement of air in the vertical dimension, is enhanced by vertical temperature differences. The steeper the temperature gradient and the larger the vertical air column in which the mixing takes place, the more vigorous the convective and turbulent mixing of the atmosphere.

## 15.2.5  TOPOGRAPHY

On a local scale, topography may affect air motion. In the United States, most large urban centers are located along sea and lake coastal areas. Contained within these large urban centers is much heavy industry. Local air flow patterns in these urban centers have a significant impact on pollution dispersion processes. Topographic features also affect local weather patterns, especially in large urban centers located near lakes, seas, and open land. Breezes from these features affect vertical mixing and pollutant dispersal. Seasonal differences in the heating and cooling of land and water surfaces may also precipitate the formation of inversions near the sea or lake shore.

River valley areas are also geographical locations that routinely suffer from industry-related pollution. Many early settlements began in river valleys because of the readily available water supply and the ease of transportation afforded to settlers by river systems within such valleys. Along with settlers came industry—the type of industry that invariably produces air pollutants. These air pollutants, because of the terrain and physical configuration of the valley, are not easily removed from the valley.

Winds that move through a typical river valley are called slope winds. Slope winds, like water, flow downhill into the valley floor. At valley floor level, slope winds transform to valley winds, which flow down the valley with the flow of the river. Down-valley winds are lighter than slope winds. The valley floor becomes flooded with a large volume of air, which intensifies the surface inversion normally produced by radiative cooling. As the inversion deepens over the course of the night, it often reaches its maximum depth just before sunrise, with the height of the inversion layer dependent on the depth of the valley and the intensity of the radiative cooling process.

Hills and mountains can also affect local air flow. These natural topographical features tend to decrease wind speed (because of their surface roughness) and form physical barriers, preventing air movement.

## 15.2.6  TEMPERATURE INVERSIONS

Temperature inversions (extreme cases of atmospheric stability) create a virtual lid on the upward movement of atmospheric pollution. Two types of inversions are important from an air quality standpoint: radiation and subsidence inversions.

*Radiation inversions* prompt the formation of fog and simultaneously trap gases and particulates, creating a concentration of pollutants. They are characteristically a nocturnal phenomenon caused by cooling of the Earth's surface. On a cloudy night, the Earth's radiant heat tends to be absorbed by water vapor in the atmosphere. Some of this is reradiated back to the surface. However, on clear winter nights, the surface more readily radiates energy to the atmosphere and beyond, allowing the ground to cool more rapidly. The air in contact with the cooler ground also cools, and the air just above the ground becomes cooler than the air above it, creating an inversion close to the ground, lasting for only a matter of hours. These radiation inversions usually begin to form at the worst time of the day for human concerns in large urban areas—during the late afternoon rush hour, trapping automobile exhaust at ground level and causing elevated concentrations of pollution for commuters. During evening hours, photochemical reactions cannot take place, so the biggest problem can be the accumulation of carbon monoxide. At sunrise, the sun warms the ground and the inversion begins to break up. Pollutants that have been trapped in the stable air mass are suddenly brought back to earth in a process known as *fumigation,* which can cause a short-lived, high concentration of pollution at ground level (Masters, 1991).

The second type of inversion is the *subsidence inversion,* usually associated with a high-pressure system. Known as *anticyclones,* they may significantly affect the dispersion of pollutants over large regions. A subsidence inversion is caused by the characteristic sinking motion of air in a high-pressure cell. Air in the middle of a high-pressure zone descends slowly. As the air descends, it is compressed and heated. It forms a blanket of warm air over the cooler air below, thus creating an inversion (located anywhere from several hundred meters above the surface to several thousand meters) that prevents further vertical movement of air.

## 15.2.7 PLUME RISE

One way to quickly determine the stability of the lower atmosphere is to view the shape of a smoke trail or plume from a tall stack located on flat terrain (see Figure 15.2). Visible plumes usually consist of pollutants emitted from a smokestack into the atmosphere. The formation and fate of the plume itself depend on a number of related factors: (1) the nature of the pollutants, (2) meteorological factors, (3) source obstructions, and (4) local topography, especially downwind. Overall, maximum ground-level concentrations will occur in a range from the vicinity of the smokestack to some distance downwind.

Figure 15.2 shows the classic types of plume behavior. When the atmosphere is slightly stable, a typical plume "cones," as indicated in Figure 15.2(a). When the atmosphere is highly unstable, a "looping" plume like the one shown in 15.2(b) forms. In the looping plume, the stream of emitted pollutants undergoes rapid mixing, and the wind causes large eddies, which may carry the entire plume down to the ground, causing high concentrations

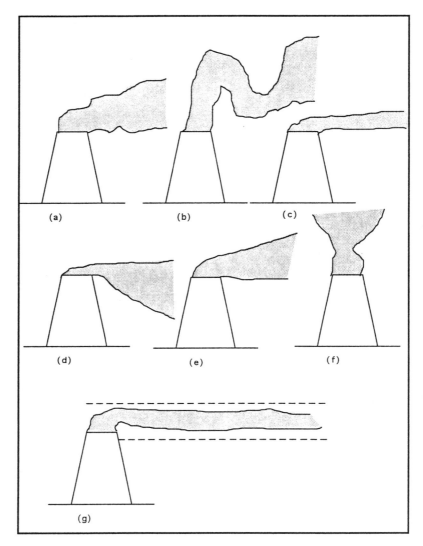

**Figure 15.2** Seven types of classic plume behavior.

close to the stack before dispersion is complete. In an extremely stable atmosphere, a "fanning" plume spreads horizontally [Figure 15.2(c)], with little mixing. When an inversion layer occurs a short distance above the plume source, the plume is said to be "fumigating" [Figure 15.2(d)]. When inversion conditions exist below the plume source, the plume is said to be "lofting" [(Figure 15.2(e)]. When conditions are neutral, the plume issuing from a smokestack tends to rise directly into the atmosphere [Figure 15.2(f)]. When an inversion layer prevails both above and below the plume source, the plume issuing from a smokestack tends to be "trapped" [Figure 15.2(g)].

Pollutants however, rarely come from a single point source (smokestack plume). In large urban areas many plumes are generated and collectively combine into a large plume (city plume) whose dispersion represents a huge environmental challenge: the high pollutant concentrations from the city plume frequently affect human health and welfare.

Air quality problems associated with dispersion of city plumes are compounded by the presence of an already contaminated environment. Even though the conventional processes that normally work to disperse emissions from point sources do occur within the city plume, because of microclimates within the city and the volume of pollutants they must handle, the conventional processes often cannot disperse pollutants effectively. Other compounding conditions present in areas where city plumes are generated (topographical barriers, surface inversions, and stagnating anticyclones) work to intensify the city plume and result in high pollutant concentrations.

## 15.2.8 TRANSPORT

Those people living east of the Mississippi River would be surprised to find out that they are breathing air contaminated by pollutants from various sources many miles from their location. Most people view pollution under the old cliché "out of sight—out of mind." As far as they are concerned, if they don't see it, it doesn't exist. For example, assume that a person on a farm heaps together a huge pile of assorted rubbish to be burned. The person preparing this bonfire probably gives little thought about the long-range transport (and consequences) of any contaminants that might be generated from that bonfire. This person simply has trash he or she no longer wants, and an easy solution is to burn it.

This pile of rubbish may contain various objects: discarded rubber tires, old compressed-gas bottles, assorted plastic containers, paper, oils and greases, wood, and old paint cans. The person burning it doesn't consider this hazardous material, though, just household trash. When the pile of rubbish burns, a huge plume of smoke forms and is carried away by a westerly wind. The person looks downwind and notices that the smoke disappears just a few miles over the property line. The dilution processes and the enormity of the atmosphere work together to dissipate and move away the smoke plume; the fire starter doesn't give it a second thought. However, elevated levels of pollutants from many such fires may occur hundreds to thousands of miles downwind of the combination point sources producing such plumes. The result is that people living many miles from such pollution generators end up breathing contaminated air, transported over distance to their location.

## 15.3 DISPERSION MODELS

Air quality models are used to predict or describe the fate of airborne gases, particulate matter, and ground-level concentrations downwind of point sources. To determine the significance of air quality impact to a particular area, the first consideration is normal background concentrations—those pollutant concentrations from natural sources and/or distant, unidentified man-made sources. Each particular geographical area has a "signature," or background level of contamination, for certain pollutants. An area, for example, might normally have a particulate matter reading of 30–40 $\mu g/m^3$. If particulate matter readings are significantly higher than the background level, this suggests an additional source. To establish background contaminations for a particular source, air quality data related to that vicinity must be collected and analyzed.

The USEPA recognized that in calculating the atmospheric dispersion of air pollutants, some means by which consistency could be maintained in air quality analysis had to be established. Thus, the USEPA promulgated two guidebooks to assist in modeling for air quality analyses: *Guidelines on Air Quality Models* (1978) and *Industrial Source Complex (ISC) Dispersion Models User's Guide* (1986).

In performing dispersion calculations, particularly for health effect studies, the USEPA and other recognized experts in the field recommend following a four-step procedure:

(1) Estimate the rate, duration, and location of the release into the environment.
(2) Select the best available model to perform the calculations.
(3) Perform the calculations and generate downstream concentrations, including lines of constant concentration (isopleths) resulting from the source emission(s).
(4) Determine what effect, if any, the resulting discharge has on the environment, including humans, animals, vegetation, and materials of construction. These calculations often include estimates of the so-called vulnerability zones—that is, regions that may be adversely affected because of the emissions (Holmes et al., 1993).

Before beginning any dispersion determination activity, you must first determine the acceptable ground-level concentration of the waste pollutant(s). Local meteorological conduits and local topography must be considered, and having an accurate knowledge of the constituents of the waste gas and its chemical and physical properties is paramount.

Air quality models provide a relatively inexpensive means of determining compliance and predicting the degree of emission reduction necessary to attain ambient air quality standards. Under the 1977 Clean Air Act Amendments, the use of models is required for the evaluation of permit applications associated with permissible increments under the so-called Prevention of Significant Deterioration (PSD) requirements, which require localities "to protect and enhance" air that is not contaminated (Godish, 1997).

Several dispersion models have been developed. These models are mathematical descriptions (equations) of the meteorological transport and dispersion of air contaminants in a particular area, which permit estimates of contaminant concentrations in plumes from either a ground-level or an elevated source (Carson & Moses, 1969). User-friendly modeling programs are available now that produce quick, accurate results from the operator's pertinent data.

This chapter's intent is not to develop each dispersion model in detail but rather to recommend the one with the greatest applicability today. Probably the best atmospheric dispersion workbook for modeling published to date is that by Turner (1970) for the USEPA, and most of the air dispersion models used today are based on the Pasquill-Gifford Model.

## 15.4  MAJOR AIR POLLUTANTS

The most common and widespread anthropogenic pollutants currently emitted are sulfur dioxide ($SO_2$), nitrogen oxides ($NO_x$), carbon monoxide (CO), carbon dioxide ($CO_2$), volatile organic compounds (hydrocarbons), particulates, lead, and several toxic chemicals. Table 15.1 lists important air pollutants and their sources.

### 15.4.1  NATIONAL AMBIENT AIR QUALITY STANDARDS (NAAQS)

Recall that in the United States, the Environmental Protection Agency (EPA) regulates air quality under the Clean Air Act (CAA) and amendments, which charged the federal government with developing uniform National

TABLE 15.1.

| Pollutant | Source |
|---|---|
| Sulfur and nitrogen oxides | From fossil fuel combustion |
| Carbon monoxide | Mostly from motor vehicles |
| Volatile organic compounds | From vehicles and industry |
| Ozone | From atmospheric reactions between nitrogen oxides and organic compounds |

*Source:* USEPA, *Environmental Progress and Challenges,* p. 13, 1988.

Ambient Air Quality Standards (NAAQS) (discussed in Chapter 14). These include a dual standard requirement of primary standards (covering criteria pollutants) designed to protect health and secondary standards to protect public welfare. Primary standards were to be achieved by July 1975, and secondary standards in "a reasonable period of time." Pollutant levels protective of public welfare take priority (and are more stringent) than those for public health; achievement of the primary health standard had immediate priority. In 1971 the USEPA promulgated NAAQS for six classes of air pollutants. Later, in 1978, an air quality standard was also promulgated for lead, and the photochemical oxidant standard was revised to an ozone ($O_3$) standard (the ozone permissible level was increased). The PM standard was revised and redesignated $PM_{10}$ standard in 1987. This revision reflected the need for a PM standard based on particle sizes ($\leq 10$ μm) that have the potential for entering the respiratory tract and affecting human health. The National Ambient Air Quality Standards (with 1998 updates) are summarized in Table 14.1.

Thus, air pollutants were categorized into two groups: primary and secondary. Primary pollutants are emitted directly into the atmosphere where they exert an adverse influence on human health or the environment. Of particular concern are primary pollutants emitted in large quantities: carbon dioxide, carbon monoxide, sulfur dioxide, nitrogen dioxides, hydrocarbons, and particulate matter (PM). Once in the atmosphere, primary pollutants may react with other primary pollutants or atmospheric compounds such as water vapor to form secondary pollutants. A secondary pollutant that has received much attention is acid precipitation, which is formed when sulfur or nitrogen oxides react with water vapor in the atmosphere.

## 15.4.2  SULFUR DIOXIDE ($SO_2$)

Sulfur enters the atmosphere in the form of corrosive sulfur dioxide ($SO_2$) gas, a colorless gas possessing the sharp, pungent odor of burning rubber. On a global basis, nature and anthropogenic activities produce sulfur dioxide in roughly equivalent amounts. Its natural sources include volcanoes, decaying organic matter, and sea spray, while anthropogenic sources include combustion of sulfur-containing coal and petroleum products and smelting of nonferrous ores. According to the World Resources Institute and International Institute for Environment and Development (WRI & IIED, 1988–89), in industrial areas much more sulfur dioxide comes from human activities than from natural sources. Sulfur-containing substances are often present in fossil fuels; $SO_2$ is a product of combustion that results from the burning of sulfur-containing materials. The largest single source of sulfur dioxide is from the burning of fossil fuels to generate electricity. Thus, near major industrialized areas, it is often encountered as an air pollutant.

In the air, sulfur dioxide converts to sulfur trioxide ($SO_3$) and sulfate particles ($SO_4$). Sulfate particles restrict visibility, and in the presence of water form sulfur acid ($H_2SO_4$), a highly corrosive substance that also lowers visibility. According to MacKenzie and El-Ashry (1988), global output of sulfur dioxide has increased sixfold since 1900. Most industrial nations, however, since 1975–1985, have lowered sulfur dioxide levels by 20% to 60% by shifting away from heavy industry and imposing stricter emission standards. Major sulfur dioxide reductions have come from burning coal with a lower sulfur content and from using less coal to generate electricity.

Two major environmental problems have developed in highly industrialized regions of the world where the atmospheric sulfur dioxide concentration has been relatively high: sulfurous smog and acid rain. Sulfurous smog is the haze that develops in the atmosphere when molecules of sulfuric acid accumulate, growing in size as droplets until they become sufficiently large to serve as light scatterers. The second problem, acid rain, is precipitation contaminated with dissolved acids such as sulfuric acid. Acid rain has posed a threat to the environment by causing certain lakes to become devoid of aquatic life.

## 15.4.3 NITROGEN OXIDES ($NO_x$)

Seven oxides of nitrogen are known to occur—$NO$, $NO_2$, $NO_3$, $N_2O$, $N_2O_3$, $N_2O_4$, and $N_2O_5$—but only two are important in the study of air pollution: nitric oxide ($NO$) and nitrogen dioxide ($NO_2$). Nitric oxide is produced by both natural and human actions. Soil bacteria are responsible for the production of most of the nitric oxide produced naturally and released to the atmosphere. Within the atmosphere, nitric oxide readily combines with oxygen to form nitrogen dioxide, and together, those two oxides of nitrogen are usually referred to as $NO_x$ (nitrogen oxides). $NO_x$ is formed naturally by lightning and by decomposing organic matter. Approximately 50% of anthropogenic $NO_x$ is emitted by motor vehicles, and about 30% comes from power plants, with the other 20% produced by industrial processes.

Scientists distinguish between two types of $NO_x$—thermal and fuel—depending on its mode of formation. Thermal $NO_x$ is created when nitrogen and oxygen in the combustion air (such as those within internal combustion engines) are heated to a high enough temperature (above 1000 K) to cause nitrogen ($N_2$) and oxygen ($O_2$) in the air to combine. Fuel $NO_x$ results from the oxidation (i.e., combines with oxygen in the air) of nitrogen contained within a fuel such as coal. Both types of $NO_x$ generate nitric oxide first, and then when vented and cooled, a portion of nitric oxide is converted to nitrogen dioxide. Although both thermal and fuel $NO_x$ can be significant contributors to the total $NO_x$ emissions, fuel $NO_x$ is usually the dominant source, with approximately 50% coming from power plants (stationary sources) and the other half released by automobiles (mobile sources).

Nitrogen dioxide is about four times more toxic than nitric oxide, and is a much more serious air pollutant. Nitrogen dioxide, at high concentrations, is believed to contribute to heart, lung, liver, and kidney damage. In addition, because nitrogen dioxide occurs as a brownish haze (giving smog its reddish-brown color), it reduces visibility. When nitrogen dioxide combines with water vapor in the atmosphere, it forms nitric acid ($HNO_3$), a corrosive substance that, when precipitated out as acid rain, causes damage to plants and corrosion of metal surfaces.

$NO_x$ levels rose in several countries and then leveled off or declined during the 1970s. During this same time frame (see Figure 15.3), levels of nitrogen oxide have not dropped as dramatically as those of sulfur dioxide, primarily because a large part of total $NO_x$ emissions comes from millions of motor vehicles, while most sulfur dioxide is released by a relatively small number of emission-controlled, coal-burning power plants.

### 15.4.4  CARBON MONOXIDE (CO)

Carbon monoxide is a colorless, odorless, tasteless gas that is by far the most abundant of the primary pollutants, as Table 15.2 indicates. When inhaled, carbon monoxide gas restricts the blood's ability to absorb oxygen,

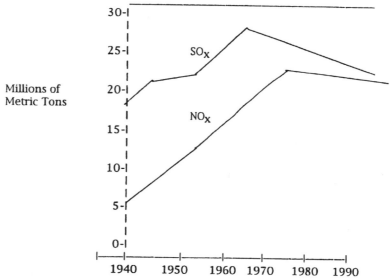

**Figure 15.3** Annual emissions of $SO_x$ and $NO_x$ in the United States, 1940–1987. (*Source:* USEPA, *National Air Pollutant Emissions Estimates,* Washington, D.C., 1989.)

TABLE 15.2. United States Emission Estimates, 1986 ($10^{12}$ g/yr).

| Source | $SO_x$ | $NO_x$ | VOC | CO | Lead | PM |
|---|---|---|---|---|---|---|
| Transportation | 0.9 | 8.5 | 6.5 | 42.6 | 0.0035 | 1.4 |
| Stationary source fuel | 17.2 | 10.0 | 2.3 | 7.2 | 0.0005 | 1.8 |
| Industrial processes | 3.1 | 0.6 | 7.9 | 4.5 | 0.0019 | 2.5 |
| Solid waste disposal | 0.0 | 0.1 | 0.6 | 1.7 | 0.0027 | 0.3 |
| Miscellaneous | 0.0 | 0.1 | 2.2 | 5.0 | 0.0000 | 0.8 |
| Total | 21.2 | 19.3 | 19.5 | 61.0 | 0.0086 | 6.8 |

Source: USEPA, *National Air Pollutant Emission Estimates 1940–1986*, Washington, D.C., 1988.

causing angina, impaired vision, and poor coordination. Carbon monoxide has little direct effect on ecosystems but has an indirect environmental impact by contributing to the greenhouse effect and depletion of the Earth's protective ozone layer.

The most important natural source of atmospheric carbon monoxide is the combination of oxygen with methane ($CH_4$), a product of the anaerobic decay of vegetation. (Anaerobic decay takes place in the absence of oxygen.) At the same time, however, carbon monoxide is removed from the atmosphere by the activities of certain soil microorganisms, so the net result is a harmless average concentration that is less than 0.12–15 ppm in the northern hemisphere. Because stationary source combustion facilities are under much tighter environmental control than are mobile sources, the principal source of carbon monoxide caused by human activities is motor vehicle exhaust, which contributes about 70% of all CO emissions in the United States (see Figure 15.4).

## 15.4.5 VOLATILE ORGANIC COMPOUNDS

Volatile organic compounds (VOCs) (also listed under the general heading of hydrocarbons) encompass a wide variety of chemicals that exclusively contain hydrogen and carbon. Emissions of volatile hydrocarbons from human resources are primarily the result of incomplete combustion of fossil fuels. Fires and the decomposition of matter are the natural sources. Of the VOCs that occur naturally in the atmosphere, methane ($CH_4$) is present at highest concentrations (approximately 1.5 ppm). But even at relatively high concentrations, methane does not interact chemically with other substances and causes no ill health effects. However, in the lower atmosphere, sunlight causes VOCs to combine with other gases such as $NO_2$, oxygen, and CO to form secondary pollutants such as formaldehyde, ketones, ozone, peroxyacetyl nitrate (PAN), and other types of photochemical oxidants. These active chemicals irritate the eyes, damage respiratory systems, and damage vegetation.

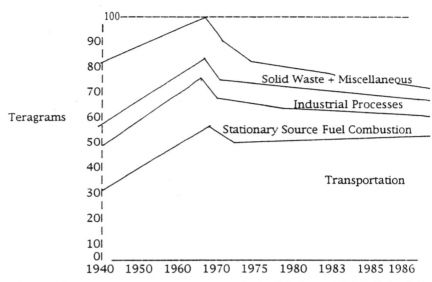

**Figure 15.4** Trends in carbon monoxide emissions, 1940–1986. (*Source:* USEPA, *National Air Pollutant Emission Estimates 1940–1986,* Washington, D.C., 1988.)

### 15.4.6 OZONE AND PHOTOCHEMICAL SMOG

By far the most damaging photochemical air pollutant is ozone (each ozone molecule contains three atoms of oxygen and thus is written $O_3$). Other photochemical oxidants [peroxyacetyl nitrate (PAN), hydrogen peroxide ($H_2O_2$), and aldehydes] play minor roles. All of these are secondary pollutants because they are not emitted but are formed in the atmosphere by photochemical reactions involving sunlight and emitted gases, especially $NO_x$ and hydrocarbons.

Ozone is a bluish gas, about 1.6 times heavier than air, and relatively reactive as an oxidant. Ozone is present in a relatively large concentration in the stratosphere and is formed naturally by ultraviolet radiation. At ground level, ozone is a serious air pollutant; it has caused serious air pollution problems throughout the industrialized world, posing threats to human health and damaging foliage and building materials.

According to MacKenzie and El-Ashry (1988), ozone concentrations in industrialized countries of North America and Europe are up to three times higher than the level at which damage to crops and vegetation begins. Ozone harms vegetation by damaging plant tissues, inhibiting photosynthesis, and increasing susceptibility to disease, drought, and other air pollutants.

In the upper atmosphere, where "good" (vital) ozone is produced, ozone is being depleted by anthropogenic emission of chemicals on the ground. With this increase, concern has been raised over a potential upset of the dynamic equilibria among stratospheric ozone reactions, with a consequent reduction in ozone concentration. This is a serious situation, because stratospheric ozone absorbs much of the incoming solar ultraviolet (UV) radiation. As a UV shield, ozone helps to protect organisms on the Earth's surface from some of the harmful effects of this high-energy radiation. If not interrupted, UV radiation could cause serious damage by disruption of genetic material, which could lead to increased rates of skin cancers and inheritable problems.

In the mid-1980s a serious problem with ozone depletion became apparent. A springtime decrease in the concentration of stratospheric ozone (ozone holes) had been observed at high latitudes, most notably over Antarctica between September and November. Scientists strongly suspect that chlorine atoms or simple chlorine compounds may play a key role in this ozone depletion problem.

On rare occasions, it is possible for upper stratospheric ozone (good ozone) to enter the lower atmosphere (troposphere). Generally, this phenomenon only occurs during an event of great turbulence in the upper atmosphere. On rare incursions, atmospheric ozone reaches ground level for a short period of time. Most of the tropospheric ozone is formed and consumed by endogenous photochemical reactions, which are the result of the interaction of hydrocarbons, oxides of nitrogen, and sunlight, which produces a yellowish-brown haze commonly called smog (Los Angeles-type smog).

Although the incursion of stratospheric ozone into the troposphere can cause smog formation, the actual formation of Los Angeles-type smog involves a complex group of photochemical interactions. These interactions are between anthropogenically emitted pollutants (NO and hydrocarbons) and secondarily produced chemicals (PAN, aldehydes, $NO_2$, and ozone). Note that the concentrations of these chemicals exhibit a pronounced diurnal pattern, depending on their rate of emission and on the intensity of solar radiation and atmospheric stability at different times of the day (Freedman, 1989). This pattern is illustrated in Figure 15.5 for the important pollutant gases that contribute to Los Angeles-type smog.

If we look at Figure 15.5 and follow the time line for the presence of various air pollutants in the atmosphere of Los Angeles, it is obvious that NO (emitted as $NO_x$) has a morning peak concentration at 0600–0700, largely caused by emissions from morning rush-hour vehicles. Hydrocarbons are emitted both from vehicles and refineries; they display a similar pattern to that of NO, except that their peak concentration is slightly later. In bright sunlight, the NO is photochemically oxidized to $NO_2$, resulting in a decrease in NO concentration and a peak of $NO_2$ at 0700–0900. Photochemical reactions involving $NO_2$ produce O atoms, which react with $O_2$ to form $O_3$. These result in a net decrease in $NO_2$ concentration and an increase in $O_3$ concentration, peaking between 1200 and 1500. Aldehydes, also formed photochemically, peak earlier than $O_3$. As the day proceeds, the various gases decrease in concentration as they are diluted by fresh air masses or are consumed by photochemical reactions. This cycle is typical of an area that experiences photochemical smog and is repeated daily (Urone, 1976)

A tropospheric ozone budget for the northern hemisphere is shown in Table 15.3. The considerable range of the estimates reflects uncertainty in the calculation of the ozone fluxes. On average, stratospheric incursions account for about 18% for the total ozone influx to the troposphere, while endogenous photochemical production accounts for the remaining 82%. About 31% of the tropospheric ozone is consumed by oxidative reactions in vegetative and inorganic suffocates at ground level, while the other 69% is consumed by photochemical reactions in the atmosphere (Freedman, 1989).

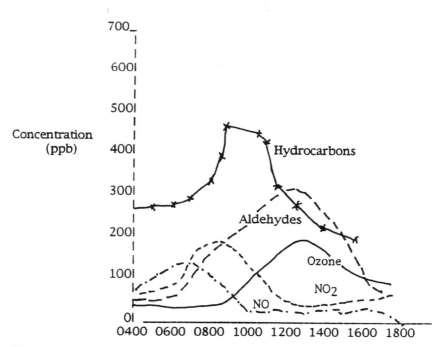

**Figure 15.5** Average concentration of various air pollutants in the atmosphere of Los Angeles during a day of eye irritation. (*Source:* Adaptation from Haagen-Smit, A. J. & Wayne, L. G., "Atmospheric reactions and scavenging processes." Air Pollution, (A. C. Stern, ed.). 3rd ed., Vol. 1, pp. 235–288. New York: Academic Press, 1976.)

TABLE 15.3. Tropospheric Ozone Budget (Northern Hemisphere) (kg/ha-year).

| | |
|---|---|
| Transport from stratosphere | 13–20 |
| Photochemical production | 48–78 |
| Destruction at ground | 18–35 |
| Photochemical destruction | 48–55 |

*Source:* Adaptation from Hov, O. (1984). "Ozone in the troposphere: High level pollution." *Ambio* 13: 73–79.

## 15.4.7 CARBON DIOXIDE

Carbon-laden fuels, when burned, release carbon dioxide ($CO_2$) into the atmosphere. Much of this carbon dioxide is dissipated and then absorbed by ocean water; some is taken up by vegetation through photosynthesis, and some remains in the atmosphere. Today, the concentration of carbon dioxide in the atmosphere is approximately 350 ppm, and is rising at a rate of approximately 20 ppm every decade. The increasing rate of combustion of coal and oil has been primarily responsible for this occurrence, which may eventually have an impact on global climate.

## 15.4.8 PARTICULATE MATTER

Atmospheric particulate matter is defined as any dispersed matter, solid or liquid, in which the individual aggregates are larger than single small molecules, but smaller than about 500 μm. Particulate matter is extremely diverse and complex, because size and chemical composition (as well as atmospheric concentrations) are important characteristics (Masters, 1991).

A number of terms are used to categorize particulates, depending on their size and phase (liquid or solid). These terms are listed and described in Table 15.4.

Dust, spray, forest fires, and the burning of certain types of fuels are among the sources of particulates in the atmosphere. Even with the implementation of stringent emission controls (which have worked to reduce particulates in the atmosphere), the U.S. Office of Technology Assessment (Postel, 1987) estimates that current levels of particulates and sulfates in ambient air may cause the premature death of 50,000 Americans every year.

## 15.4.9 LEAD

Lead is emitted to the atmosphere primarily from human sources (such as burning leaded gasoline) in the form of inorganic particulates. In high concentrations, lead can damage human health and the environment. Once lead enters an ecosystem, it remains there permanently. In humans and

TABLE 15.4. **Atmospheric Particulate.**

| Term | Description |
|------|-------------|
| Aerosol | General term for particles suspended in air |
| Mist | Aerosol consisting of liquid droplets |
| Dust | Aerosol consisting of solid particles that are blown into the air or are produced from larger particles by grinding them down |
| Smoke | Aerosol consisting of solid particles or a mixture of solid and liquid particles produced by chemical reactions such as fires |
| Fume | Generally means the same as smoke, but often applies specifically to aerosols produced by condensation of hot vapors, especially of metals |
| Plume | The geometrical shape or form of the smoke coming out of a stack or chimney |
| Fog | Aerosol consisting of water droplets |
| Haze | Any aerosol, other than fog, that obscures the view through the atmosphere |
| Smog | Popular term originating in England to describe a mixture of smoke and fog; implies photochemical pollution |

animals, lead can affect the neurological system and cause kidney disease. In plants, lead can inhibit respiration and photosynthesis as well as block the decomposition of microorganisms. Since the 1970s, stricter emission standards have caused a dramatic reduction in lead output.

## 15.5 SUMMARY

The problems of air quality and pollution are varied, many, and complex. We know, intellectually, how critical they are to our health and well-being, to all aspects of our lives, but not only is the information we have at hand easy to ignore (unless you live in a community with serious pollution problems), the information is sadly incomplete.

What problems with our air and atmosphere are anthropogenically generated? What situations are the result of long-term natural cycles? Which aspects can we affect, and which ones do we have no control over? While our uncertainty of the best course of action to take is no reason to halt study or legislation to control emissions or concern on these issues, only time will provide the answers.

## 15.6 REFERENCES

Barker, J. R. & Tingey, D. T., *Air Pollution Effects on Biodiversity*. New York: Van Nostrand Reinhold, 1991.

Carson, J. E. & Moses, H., "The validity of several plume rise formulas," *J. Air Pol. Cont. Assoc.*, 19(11):862 (1969).

Freedman, B., *Environmental Ecology*. New York: Academic Press, 1989.

Godish, T., *Air Quality*. 3rd ed., Boca Raton, FL: Lewis Publishers, 1997.

Holmes, G., Singh, B. R., & Theodore, L., *Handbook of Environmental Management & Technology*. New York: John Wiley, 1993.

MacKenzie, J. J. & El-Ashry, T., *Ill Winds: Airborne Pollutant's Toll on Trees and Crops*. Washington, D.C.: World Resource Institute, 1988.

Masters, G. M., *Introduction to Environmental Engineering and Science*, Englewood Cliffs, NJ: Prentice Hall, 1991.

Peavy, H. S., Rowe, D. R., & Tchobanglous, G., *Environmental Engineering*. New York: McGraw-Hill, 1985.

Postel, S., "Stabilizing Chemical Cycles," Lester R. Brown (ed.), *State of the World*. New York: Norton, 1987.

Turner, D. B., *Workbook of Atmospheric Dispersion Estimates*. Washington, D.C.: EPA, 1970.

Urone, P., "The primary air pollutants—gaseous. Their occurrence, sources, and effects." *Air Pollution*, (A. C. Stern, ed.), Vol. 1, New York: Academic Press, 1976.

WRI & IIED, *World Resources 1988–1989*. New York: Basic Books, 1988.

# CHAPTER 16

# Air Pollution Control Technology

*There are two primary motivations behind the utilization of industrial air pollution control technologies. These are*

*1. They must be used because of legal or regulatory requirements*
*2. They are integral to the economical operation of an industrial process*

*Although economists would point out that both of these motivations are really the same, that is, it is less expensive for an industrial user to operate with air pollution control than without, the distinction in application type is an important one. . . . In general, air pollution control is used to describe those applications that are driven by regulations and/or health considerations, while applications that deal with product recovery are considered process applications. Nevertheless, the technical issues, equipment design, operation, etc., will be similar if not identical. In fact, what differs between these uses is that the economics that affect the decision making process will often vary to some degree. (Heumann, 1997, p. xv)*

## 16.1 INTRODUCTION

CHAPTERS 13 through 15 set the foundation for the discussion presented in this chapter. Now that you have a clear picture of the problems that air pollution control is trying to solve, the time has come to examine the measures used to control it.

Two important factors related to the topic are presented in the opening sentence of this chapter's introductory quote: control technology and regulation. Neither is more important than the other. In fact, in many ways, they drive each other.

Air pollution control begins with regulation. Regulations (for example, to clean up, reduce, or eliminate a pollutant emission source) in turn are generated because of certain community concerns. Buonicore et al. (1992) point out that regulations usually evolve around three considerations:

(1) Legal limitations imposed for the protection of public health and welfare
(2) Social limitations imposed by the community in which the pollution source is or will be located
(3) Economic limitations imposed by marketplace constraints

The engineer assigned to mitigate an air pollution problem must ensure that the design control methodology used will bring the source into full compliance with applicable regulations. To accomplish this feat, environmental engineers must first understand the problem(s) and then rely heavily on technology to correct the situation. Various air pollution control technologies are available to environmental engineers or scientists working to mitigate air pollution source problems. By analyzing the problem carefully and applying the most effective method for the situation, the engineer or scientist can ensure that a particular pollution source is brought under control and the responsible parties are in full compliance with regulations.

In this chapter, we discuss the various air pollution control technologies available to environmental scientists and engineers in mitigating air pollution source problems.

## 16.2 AIR POLLUTION CONTROL: CHOICES

Assuming that the design engineer has a complete knowledge of the contaminant and the source, all available physical and chemical data on the effluent from the source, and the regulations of the control agencies involved, he or she must then decide which control methodology to employ. Since only a few control methods exist, the choice is limited. Control of atmospheric emissions from a process will generally consist of one of four methods, depending on the process, types, fuels, availability of control equipment, etc. The four general control methods are (1) elimination of the process entirely or in part, (2) modification of the operation to a fuel that will give the desired level of emission, (3) installation of control equipment between the pollutant source and the receptor, and (4) relocation of the operation.

Tremendous costs are involved with eliminating or relocating a complete process, which makes either of these choices the choice of last resort. Eliminating a process is no easy undertaking, especially when the "process" to be eliminated is the process for which the facility exists. Relocation is not always an answer either. Consider the real-life situation presented below.

---

Cedar Creek Composting (CCC) Facility was built in 1970. A 44-acre site designed to receive and process to compost wastewater biosolids from six local wastewater treatment plants, CCC composted biosolids at the rate of 17.5 dry tons per day. CCC used the aerated static pile (ASP) method to produce pathogen-free, humus-like material that could be beneficially used as an organic soil amendment. The final compost product was successfully marketed under a registered trademark name.

Today, Cedar Creek Composting Facility is no longer in operation. The site was shut down in early 1997. From an economic point of view, CCC was highly successful. When a pile of compost had completed the entire composting process (including curing), dumptruck after dumptruck would line the street outside the main gate, waiting in the hope of buying loads of the popular product. In fact, CCC could not produce compost fast enough to satisfy the demand. No, economics was not the problem.

What was the problem? The answer to this is actually twofold. The first problem was social limitations imposed by the community where the compost site was located. In 1970, the 44 acres CCC occupied were located in a rural area. CCC's only neighbor was a regional, small airport on its eastern border. CCC was completely surrounded by woods on the other three sides. The nearest town was two miles away. But by the mid-1970s, things started to change. Population growth and its accompanying urban sprawl quickly turned forested lands into housing complexes and shopping centers. CCC's western border soon became the site of a two-lane road that was upgraded to four and then to six lanes. CCC's northern fence separated it from a mega-shopping mall. On the southern end of the facility, acres of houses, playgrounds, swimming pools, tennis courts, and a golf course were built. CCC became an island surrounded by urban growth. Further complicating the situation was the airport; it expanded to the point that, by 1985, three major airlines used the facility.

CCC's ASP composting process was not a problem before the neighbors moved in. We all know that dust and odor are not problems until there are neighbors to complain. CCC was attacked from all four sides. The first complaints came from the airport. The airport complained that dust from the static piles of compost was interfering with air traffic control.

The new, expanded highway brought several thousand new commuters right alongside CCC's western fence line. Commuters started complaining anytime the compost process was in operation; they complained primarily about the odor.

After the enormous housing project was completed and people took up residence there, complaints were raised on a daily basis. The new homeowners complained about the earthy odor and the dust that blew from the compost piles onto their properties anytime they were downwind from the site. The shoppers at the mall also complained about the odor.

City Hall received several thousand complaints over the first few months before they took any action. The city environmental engineer was told to approach CCC's management and see if some resolution of the problem could be effected. CCC management listened to the engineer's concerns but stated that there wasn't a whole lot that the site could do to rectify the problem.

As you might imagine, this was not the answer the city was hoping to get. Feeling the increasing pressure from local inhabitants, commuters, shoppers, and airport management people, the city brought the local state representatives into the situation. The two state representatives for the area immediately began a campaign to close down the CCC facility.

CCC was not powerless in this struggle—after all, CCC was there first. The developers and those people in those new houses didn't have to buy land right next to the facility—right? Besides, CCC had the USEPA on their side. CCC was taking a waste product no one wanted, one that traditionally ended up in the local landfill (taking up valuable space), and turning it into a beneficial reuse product. CCC was helping to conserve and protect the local environment, a noble endeavor.

The city politicians didn't really care about noble endeavors, but they did care about the concerns of their constituents, the voters. They continued their assault through the press, electronic media, legislatively, and by any other means they could bring to bear.

CCC management understood the problem and felt the pressure. They had to do something, and they did. Their environmental engineering division was assigned the task of coming up with a plan to mitigate not only CCC's odor problem, but also its dust problem. After several months of research and a pilot study, CCC's environmental engineering staff came up with a solution. The solution included enclosing the entire facility within a self-contained structure. The structure would be equipped with a state-of-the-art ventilation system and two-stage odor scrubbers. The engineers estimated that the odor problem could be reduced by 90% and the dust problem reduced by 98.99%. CCC management thought they had a viable solution to the problem and was willing to spend the 5.2 million dollars to retrofit the plant.

After CCC presented their mitigation plan to the city council, the council members made no comment but said that they needed time to study the plan. Three weeks later, CCC received a letter from the mayor stating that CCC's efforts to come up with a plan to mitigate the odor and dust problems at CCC were commendable and to be applauded but were unacceptable.

From the mayor's letter, CCC could see that the focus of attack had now changed from a social to a legal issue. The mayor pointed out that he and the city politicians had a legal responsibility to ensure the good health and well-being of local inhabitants and that certain legal limitations would be imposed and placed on the CCC facility to protect their health and welfare.

Compounding the problem was the airport. Airport officials also rejected CCC's plan to retrofit the compost facility. Their complaint stated that the dust generated at the compost facility was endangering flight operations, and even though the problem would be reduced substantially by engineering controls, the chance of control failure was always possible, and then an aircraft could be endangered. From the airport's point of view, this was unacceptable.

Several years went by, with local officials and CCC management contesting each other on the plight of the compost facility. In the end, CCC management decided they had to shut down its operation and move to another location, so they closed the facility.

After shutdown, CCC management staff immediately started looking for another site to build a new wastewater biosolids-to-compost facility. They are still looking. To date, their search has located several pieces of property relatively close to the city (but far enough away to preclude any dust and odor problems), but they have had problems finalizing any deal. Buying the land is not the problem—getting the required permits from various county agencies to operate the facility is. CCC officials were turned down in each case. The standard excuse? Not in my backyard. This phrase is so common now, it's usually abbreviated—NIMBY.

CCC officials are still looking for a location for their compost facility; they are not very optimistic about their chances of success in this matter.

---

The second pollution control method—modification of the operation to a fuel that will give the desired level of emission—often looks favorable to those who have weighed the high costs associated with air pollution control systems. Modifying the process to eliminate as much of the pollution problem as possible at the source is generally the first approach to be examined.

Often, the easiest way to modify a process for air pollution control is to change the fuel. If a power plant, for example, emits large quantities of sulfur dioxide and fly ash, conversion to cleaner-burning natural gas is cheaper than installing the necessary control equipment to reduce the pollutant emissions to permitted values.

Changing from one fuel to another, however, causes its own problems related to costs, availability, and competition. Today's fuel prices are high, and no one counts on the trend reversing. Finding a low-sulfur fuel isn't easy, especially since many industries own their own dedicated supplies (which are not available for use in other industries). With regulation compliance threatening everyone, everyone wants their share of any available low-cost, low-sulfur fuel. With limited supplies available, the law of supply and demand takes over and prices go up.

Some industries employ other process modification techniques. These may include evaluation of alternative manufacturing and production techniques, substitution of raw materials, and improved process control methods (Buonicore et al., 1992).

When elimination of the process entirely or in part, when relocation of the operation, or when modification of the operation to a fuel that will give the desired level of emission are not possible, the only alternative control method left is installation of control equipment between the pollutant source and the receptor. To accomplish this, the polluted carrier gas must pass through a control device or system, which collects or destroys the pollutant and releases the cleaned carrier gas to the atmosphere (Boubel et al., 1994). The rest of this chapter will focus on these air pollution control equipment devices and systems.

## 16.3  AIR POLLUTION CONTROL EQUIPMENT AND SYSTEMS

Several considerations must be factored into any selection decision for air pollution control equipment or systems. Careful consideration must be given to costs. No one ever said air pollution equipment/systems were inexpensive. Obviously, the equipment/system must be designed to comply with applicable regulatory emission limitations. The operational and maintenance record (costs of energy, labor, and repair parts) of each equipment/system must be evaluated. Remember, emission control equipment must be operated on a continual basis, without interruptions. Any interruption could be subject to severe regulatory penalty, which could again be quite costly.

Probably the major factor to consider in the equipment/system selection process is what type of pollutant or pollutant stream is under consideration. If the pollutant is conveyed in a carrier gas, for example, factors such as carrier gas pressure, temperature, viscosity, toxicity, density, humidity, corrosiveness, and inflammability must all be considered before any selection is made. Many of the general factors are listed in Table 16.1.

In addition to those factors listed in Table 16.1, process considerations

TABLE 16.1.  Factors in Selecting Air Pollution Control Equipment/Systems.

|  |
| --- |
| (1)  Best available technology (BAT) |
| (2)  Reliability |
| (3)  Lifetime and salvage value |
| (4)  Power requirements |
| (5)  Collection efficiency |
| (6)  Capital cost, including operation and maintenance costs |
| (7)  Track record of equipment/system and manufacturer |
| (8)  Space requirements |
| (9)  Power requirements |
| (10)  Availability of spare parts and manufacturer's representatives |

dealing with gas flow rate and velocity, pollutant concentration, allowable pressure drop, and the variability of gas and pollutant flow rates (including temperature) must all be considered.

The type of pollutant is also an important factor to be taken into consideration—gaseous or particulate. Certain pertinent questions must be asked and answered. If the pollutant, for example, is gaseous, how corrosive, inflammable, reactive, and toxic is it? After these factors have been evaluated, the focus shifts to the selection of the best air pollution control equipment/system—affordable, practical, and permitted by regulatory requirements—depending, of course, on the type of pollutant to be removed.

In the following sections two types of pollutants (dry particulates and gaseous pollutants) and the various air pollution control equipment/processes available for their removal are discussed.

## 16.4 REMOVAL OF DRY PARTICULATE MATTER

Constituting a major class of air pollutants, particulates have a variety of shapes and sizes, and as either liquid droplet or dry dust, they have a wide range of physical and chemical characteristics. Dry particulates are emitted from a variety of different sources, including both combustion and noncombustion sources in industry, mining, construction activities, incinerators, and internal combustion engines. Dry particulates also come from natural sources—volcanoes, forest fires, pollen, and windstorms.

All particles and particulate matter exhibit certain important characteristics, which, along with process conditions, must be considered in any engineering strategy to separate and remove them from a stream of carrier gas. Particulate size range and distribution, particle shape, corrosiveness, agglomeration tendencies, abrasiveness, toxicity, reactivity, inflammability, and hygroscopic tendencies must all be examined in light of equipment limitations.

When a flowing fluid (engineering and science applications consider both liquid and gaseous states as a fluid) approaches a stationary object such as a metal plate, a fabric thread, or a large water droplet, the fluid flow will diverge around that object. Particles in the fluid (because of inertia) will not follow stream flow exactly but will tend to continue in their original directions. If the particles have enough inertia and are located close enough to the stationary object they will collide with the object, and can be collected by it. This is an important phenomenon and is depicted in Figure 16.1.

Particles are collected by impaction, interception, and diffusion. Impaction occurs when the center of mass of a particle that is diverging from the fluid strikes a stationary object. Interception occurs when the particle's

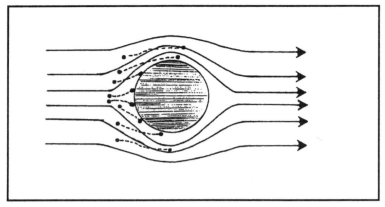

**Figure 16.1** Particle collection of a stationary object.

center of mass closely misses the object, but because of its size, the particle strikes the object. Diffusion occurs when small particulates happen to "diffuse" toward the object while passing near it. Particles that strike the object by any of these means are collected if short-range forces (chemical, electrostatic, and so forth) are strong enough to hold them to the surface (Cooper & Alley, 1990).

Different classes of particulate control equipment include gravity settlers, cyclones, electrostatic precipitators, wet (Venturi) scrubbers, and baghouses (fabric filters). In the following sections each of the major types of particulate control equipment is introduced and their advantages and disadvantages pointed out.

### 16.4.1 GRAVITY SETTLERS

Gravity settlers have long been used by industry for removing solid and liquid waste materials from gaseous streams. Simply constructed (see Figures 16.2 and 16.3), a gravity settler is actually nothing more than an enlarged chamber in which the horizontal gas velocity is slowed, allowing particles to settle out by gravity. Gravity settlers have the advantage of having low initial cost and are relatively inexpensive to operate—there's not a lot to go wrong. Although simple in design, gravity settlers require a large space for installation and have relatively low efficiency, especially for removal of small particles (<50 μm).

### 16.4.2 CYCLONE COLLECTORS

The cyclone (or centrifugal) collector removes particles by causing the entire gas stream to flow in a spiral pattern inside a tube and is the collector

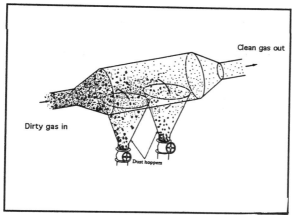

**Figure 16.2** Gravitational settling chamber. (*Source:* USEPA, *Control Techniques for Gases and Particulates,* 1971.)

of choice for removing particles greater than 10 μm in diameter. By centrifugal force, the larger particles move outward and collide with the wall of the tube. The particles slide down the wall and fall to the bottom of the cone, where they are removed. The cleaned gas flows out the top of the cyclone (see Figure 16.4). Cyclones have low construction costs and need relatively

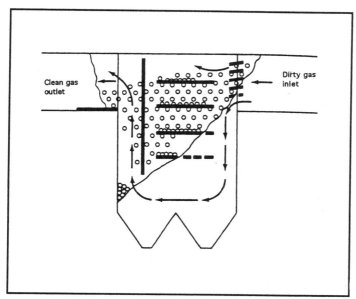

**Figure 16.3** Baffled gravitational settling chamber. (*Source:* USEPA, *Control Techniques for Gases and Particulates,* 1971.)

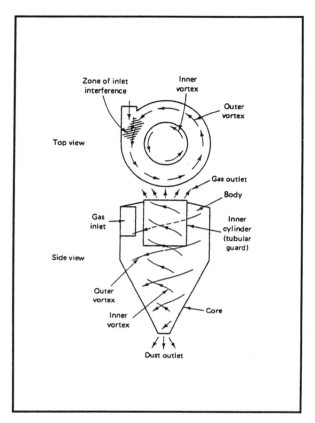

**Figure 16.4** Convection reverse-flow cyclone. (*Source:* USEPA, *Control Techniques for Gases and Particulates,* 1971.)

small space requirements for installation. However, note that the cyclone's overall particulate collection efficiency is low, especially on particles below 10 μm in size, and they do not handle sticky materials well. The most serious problems encountered with cyclones are with air flow equalization and their tendency to plug. Cyclones have been used successfully at feed and grain mills, cement plants, fertilizer plants, petroleum refineries, and other applications involving large quantities of gas containing relatively large particles.

## 16.4.3 ELECTROSTATIC PRECIPITATOR

The electrostatic precipitator is usually used to remove small particles from moving gas streams at high collection efficiencies. Widely used in power plants for removing fly ash from the gases prior to discharge, an electrostatic precipitator applies electrical force to separate particles from the gas stream. A high voltage drop is established between electrodes, and particles passing through the resulting electrical field acquire a charge. The charged particles are attracted to and collected on an oppositely charged plate, and the cleaned gas flows through the device. Periodically, the plates are cleaned by rapping to shake off the layer of dust that accumulates, and the dust is collected in hoppers at the bottom of the device (see Figure 16.5). Although electrostatic precipitators have the advantages of low operating costs, capability for operation in high-temperature applications (to 1300°F), low pressure drop, and extremely high particulate (coarse and fine) collection efficiencies, they have the disadvantages of high capital costs and space requirements.

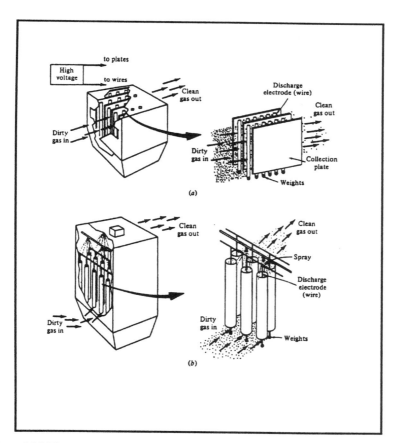

**Figure 16.5** Electrostatic precipitators: (a) plate type and (b) tube type. (*Source:* USEPA, *Control Techniques for Gases and Particulates,* 1971.)

### 16.4.4 WET (VENTURI) SCRUBBERS

Wet scrubbers (or collectors) have found widespread use in cleaning contaminated gas streams (e.g., foundry dust emissions, acid mists, and furnace fumes) because of their ability to effectively remove particulate and gaseous pollutants. Wet scrubbers vary in complexity from simple spray chambers to remove coarse particles to high efficiency systems (venturi types) to remove fine particles. Whichever system is used, operation employs the same basic principles of inertial impingement or impaction and interception of dust particles by droplets of water. The larger, heavier water droplets are easily separated from the gas by gravity. The solid particles can then be independently separated from the water, or the water can be otherwise treated before reuse or discharge. Increasing either the gas velocity or the liquid droplet velocity in a scrubber increases the efficiency because of the greater number of collisions per unit time. For the ultimate in wet scrubbing, where high collection efficiency is desired, the venturi scrubber

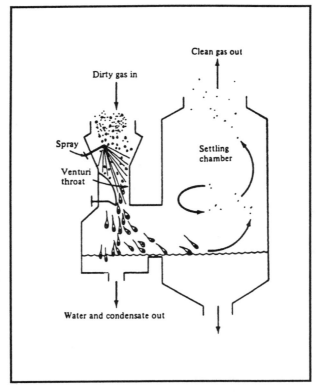

**Figure 16.6** Venturi wet scrubber. (*Source:* USEPA, *Control Techniques for Gases and Particulates,* 1971.)

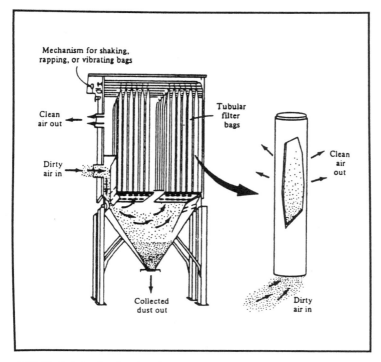

**Figure 16.7** Typical simple fabric filter baghouse design. (*Source:* USEPA, *Control Techniques for Gases and Particulates,* 1971.)

is used. The venturi operates at extremely high gas and liquid velocities with a very high pressure drop across the venturi throat. Venturi scrubbers such as the one shown in Figure 16.6 are most efficient for removing particulate matter in the size range of 0.5 to 5 μm, which makes them especially effective for the removal of submicron particulates associated with smoke and fumes.

Although wet scrubbers require relatively small space requirements, have low capital cost, and can handle high-temperature, high-humidity gas streams, their power and maintenance costs are relatively high, they may create water disposal problems, their corrosion problems are more severe than dry systems, and the final product they produce is collected wet.

## 16.4.5 BAGHOUSE (FABRIC) FILTERS

Baghouse filters (or fabric filters) are the most commonly used air pollution control filtration system. In much the same manner as the common vacuum cleaner, fabric filter material, capable of removing most particles as small as 0.5 μm and substantial quantities of particles as small as 0.1 μm, is formed into cylindrical or envelope bags and suspended in the baghouse (see Figure 16.7). The particulate-laden gas stream is forced through the fabric filter, and as the air passes through the fabric, particulates accumulate on the cloth, providing a cleaned airstream. As particulates build up on the inside surfaces of the bags, the pressure drop increases. Before the pressure drop becomes too severe, the bags must be relieved of some of the particulate layer. The particulates are periodically removed from the cloth by shaking or by reversing the air flow.

Fabric filters are relatively simple to operate, provide high overall collection efficiencies up to 99+%, and are very effective in controlling submicrometer particles, but they do have limitations. These include relatively high capital costs, high maintenance requirements (bag replacement, etc.), high space requirements, and flammability hazards for some dusts.

## 16.5 REMOVAL OF GASEOUS POLLUTANTS: STATIONARY SOURCES

In the removal of gaseous air pollutants, the principal gases of concern are the sulfur oxides ($SO_x$), carbon oxides ($CO_x$), nitrogen oxides ($NO_x$), organic and inorganic acid gases, and hydrocarbons (HC). Four major treatment processes are currently available for control of these and other gaseous emissions: absorption, adsorption, condensation, and combustion (incineration).

The decision of which single or combined air pollution control technique to use for stationary sources is not always easy. Gaseous pollutants can be controlled by a wide variety of devices, and choosing the most cost-effective, most efficient units requires careful attention to the particular operation for which the control devices are intended. Specifically, the choice of control technology depends on the pollutant(s) to be removed, the removal efficiency required, pollutant and gas stream characteristics, and specific characteristics of the site (Peavy et al., 1985).

In making the difficult and often complex decision of which air pollution control technology to employ, it is helpful to follow guidelines based on experience and set forth by Buonicore and Davis (1992) in the prestigious engineering text *Air Pollution Engineering Manual*. Table 16.2 summarizes these.

### 16.5.1 ABSORPTION

Absorption (or scrubbing) is a major chemical engineering unit operation that involves bringing contaminated effluent gas into contact with a liquid absorbent so that one or more constituents of the effluent gas are selectively dissolved into a relatively nonvolatile liquid.

Absorption units are designed to transfer the pollutant from a gas phase to a liquid phase. The absorption unit accomplishes this by providing

TABLE 16.2. Comparison of Air Control Technologies.

| Treatment Technology | Concentration and Efficiency | Comments |
|---|---|---|
| Incineration | (<100 ppmv) 90–95% efficient<br>(>100 ppmv) 95–99% efficient | Incomplete combustion may require additional controls. |
| Carbon Adsorption | (>200 ppmv) 90+% efficiency<br>(>1000 ppmv) 95+% efficiency | Recovered organics may need additional treatment; can increase cost. |
| Absorption | (<200 ppmv) 90–95% efficiency<br>(>200 ppmv) 95+% efficiency | Can blowdown stream be accommodated at site? |
| Condensation | (>2000 ppmv) 80+% efficiency | Must have low temperature or high pressure for efficiency. |

*Note:* Typically, only incineration and absorption technologies can achieve greater than 99% gaseous pollutant removal consistently.

intimate contact between the gas and the liquid, providing optimum diffusion of the gas into the solution. The actual removal of a pollutant from the gas stream takes place in three steps: (1) diffusion of the pollutant gas to the surface of the liquid; (2) transfer across the gas–liquid interface; and (3) diffusion of the dissolved gas away from the interface into the liquid (Davis & Cornwell, 1991).

Several types of absorbers are available, including spray chambers (towers or columns), plate or tray towers, packed towers, and venturi scrubbers. Pollutant gases commonly controlled by absorption include sulfur dioxide, hydrogen sulfide, hydrogen chloride, chlorine, ammonia, and oxides of nitrogen.

The two most common absorbent units in use today are the plate and packed tower systems. Plate towers contain perforated horizontal plates or trays designed to provide large liquid–gas interfacial areas. The polluted airstream is usually introduced at one side of the bottom of the tower or column and rises up through the perforations in each plate; the rising gas prevents the liquid from draining through the openings rather than through a downpipe. During continuous operation, contact is maintained between air and liquid, allowing gaseous contaminants to be removed, with clean air emerging from the top of the tower.

The packed tower scrubbing system (see Figure 16.8) is predominately used to control gaseous pollutants in industrial applications, where it typically demonstrates a removal efficiency of 90% to 95%. Usually configured in vertical fashion (Figure 16.8) the packed tower is literally "packed" with devices (see Figure 16.9) of large surface-to-volume ratio and a large void

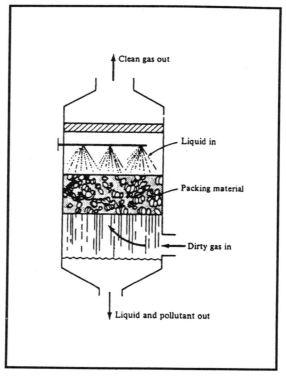

**Figure 16.8** Typical countercurrent-flow packed tower. (*Source:* USEPA, *Control Techniques for Gases and Particulates,* 1971.)

ratio that offers minimum resistance to gas flow. In addition, packing should provide even distribution of both fluid phases, be sturdy enough to support themselves in the tower, and be low-cost, available, and easily handled (Hesketh, 1991).

The flow through a packed tower is typically countercurrent, with gas entering at the bottom of the tower and liquid entering at the top. Liquid flows over the surface of the packing in a thin film, affording continuous contact with the gases.

Though highly efficient for removal of gaseous contaminants, packed towers may create liquid disposal problems, become easily clogged when gases with high particulate loads are introduced, and have relatively high maintenance costs.

## 16.5.2 ADSORPTION

Adsorption is a mass transfer process that involves passing a stream of effluent gas through the surface of prepared porous solids (adsorbents). The surfaces of the porous solid substance attract and hold the gas (the adsorbate) by either physical or chemical adsorption. In physical adsorption (a readily reversible process), a gas molecule adheres to the surface of the solid because of an imbalance of electron distribution. In chemical adsorption (not readily reversible), once the gas molecule adheres to the surface, it reacts chemically with it.

Several materials possess adsorptive properties. These materials include activated carbon, alumina, bone char, magnesia, silica gel, molecular sieves, strontium sulfate, and others. The most important adsorbent for air pollution control is activated charcoal. The surface area of activated charcoal will preferentially adsorb hydrocarbon vapors and odorous organic compounds from an airstream.

In an adsorption system (in contrast to the absorption system where the collected contaminant is continuously removed by flowing liquid), the collected contaminant remains in the adsorption bed. The most common adsorption system is the fixed-bed adsorber, which can be contained in either a vertical or horizontal cylindrical shell. The adsorbent (usually activated

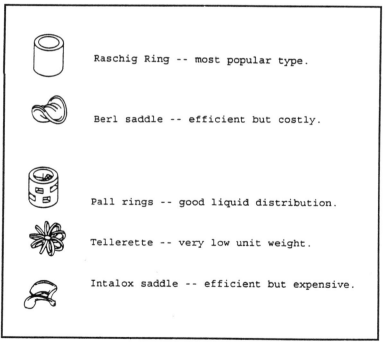

**Figure 16.9** Various packing used in packed tower scrubbers. (*Source:* Adaptation from American Industrial Hygiene Association, 1968.)

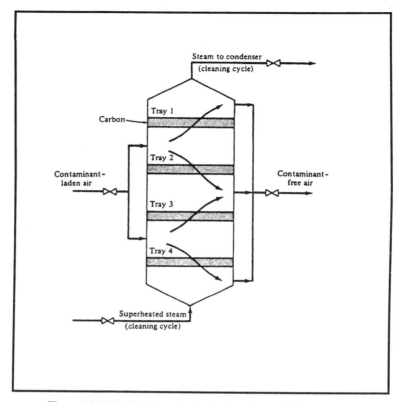

**Figure 16.10** Multiple fixed-bed adsorber. (*Source:* USEPA, 1973.)

carbon) is arranged in beds or trays in layers about 0.5 inch. thick. Multiple-beds may be arranged as shown in Figure 16.10. In multiple-bed systems, one or more beds are adsorbing vapors, while the other bed is being regenerated.

The efficiency of most adsorbers is near 100% at the beginning of the operation and remains high until a breakpoint or breakthrough occurs. When the adsorbent becomes saturated with adsorbate, contaminant begins to leak out of the bed, signaling that the adsorber should be renewed or regenerated.

Although adsorption systems are high-efficiency devices that may allow recovery of product, have excellent control and response to process changes, and have the capability of being operated unattended, they also have some disadvantages, including the need for expensive extraction schemes if product recovery is required, relatively high capital cost, and gas stream prefiltering needs (to remove any particulate capable of plugging the adsorbent bed).

### 16.5.3 CONDENSATION

Condensation is a process by which volatile gases are removed from the contaminant stream and changed into a liquid. In air pollution control, a condenser can be used in two ways: either for pretreatment to reduce the load problems with other air pollution control equipment or for effectively controlling contaminants in the form of gases and vapors.

Condensers condense vapors to a liquid phase by either increasing the system pressure without a change in temperature or by decreasing the system temperature to its saturation temperature without a pressure change. Condensation is affected by the composition of the contaminant gas stream. When different gases are present in the stream that condense under different conditions, condensation is hindered.

There are two basic types of condensation equipment—surface and contact condensers. A surface condenser is normally a shell-and-tube heat exchanger (see Figure 16.11). It uses a cooling medium of air or water where the vapor to be condensed is separated from the cooling medium by a metal wall. Coolant flows through the tubes while the vapor is passed over the tubes, condenses on the outside of the tubes and drains off to storage (USEPA, 1971).

In a contact condenser (which resembles a simple spray scrubber), the vapor is cooled by spraying liquid directly on the vapor stream (see Figure 16.12). The cooled vapor condenses, and the water and condensate mixture are removed, treated, and disposed of.

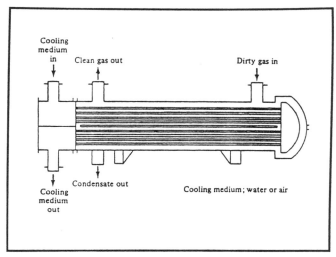

**Figure 16.11** Surface condenser. (*Source:* USEPA, *Control Techniques for Gases and Particulates,* 1971.)

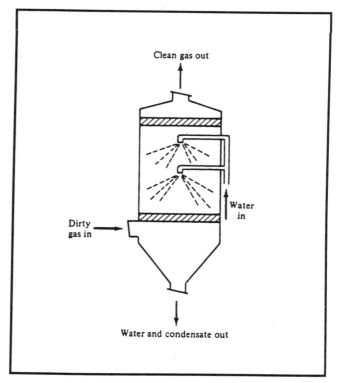

**Figure 16.12** Contact condenser. (*Source:* USEPA, *Control Techniques for Gases and Particulates,* 1971.)

In general, contact condensers are less expensive, more flexible, and simpler than surface condensers, but surface condensers require much less water and produce many times less wastewater that must be treated than do contact condensers. Condensers are used in a wide range of industrial applications, including petroleum refining, petrochemical manufacturing, basic chemical manufacturing, dry cleaning, and degreasing.

### 16.5.4 COMBUSTION

Even though combustion (or incineration) is a major source of air pollution, it is also, if properly operated, a beneficial air pollution control system in which the objective is to convert certain air contaminants (usually CO and hydrocarbons) to innocuous substances such as carbon dioxide and water (USEPA, 1973).

Combustion is a chemical process defined as rapid, high-temperature gas-phase oxidation. The combustion equipment used to control air pollution emissions is designed to push these oxidation reactions as close to complete combustion as possible, leaving a minimum of unburned residue. The operation of any combustion operation is governed by four variables: oxygen, temperature, turbulence, and time. For complete combustion to occur, oxygen must be available and put into contact with sufficient temperature (turbulence) and held at this temperature for a sufficient time. These four variables are not independent—changing one affects the entire process.

Depending upon the contaminant being oxidized, equipment used to control waste gases by combustion can be divided into three categories: direct-flame combustion (or flaring), thermal combustion (afterburners), or catalytic combustion.

### 16.5.4.1 Direct-Flame Combustion (Flaring)

Direct-flame combustion devices (flares) are the most commonly used air pollution control devices by which waste gases are burned directly (with or without the addition of a supplementary fuel). Common flares include steam-assisted, air-assisted, and pressure-head types. Flares are normally elevated from 100 to 400 feet to protect the surroundings from heat and flames. Often designed for steam injection at the flare top (see Figure 16.13), flares commonly use steam in this application because steam provides sufficient turbulence to ensure complete combustion, which prevents production of visible smoke or soot. Flares are also noisy, which can cause problems for adjacent neighborhoods, and some flares produce oxides of nitrogen, thus creating a new air pollutant. Figure 16.14 shows a steam-assisted flare system commonly used in industry.

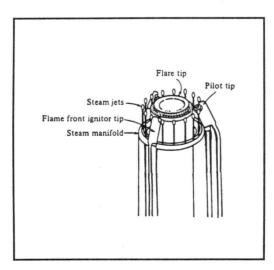

**Figure 16.13** Close-up of steam-injection type flare head. (*Source:* USEPA, *Control Techniques for Gases and Particulates,* 1971.)

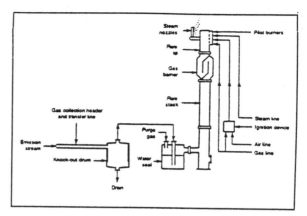

**Figure 16.14** Schematic of steam-assisted flare system. (*Source:* USEPA, 1986.)

### 16.5.4.2 Thermal Combustion (Afterburners)

The thermal incinerator or afterburner is usually the unit of choice in cases where the concentration of combustible gaseous pollutants is too low to make flaring practical. Widely used in industry, typically, the thermal combustion system operates at high temperatures. Within the thermal incinerator, the contaminant airstream passes around or through a burner and into a refractory-line residence chamber where oxidation occurs (see Figure 16.15). Flue gas from a thermal incinerator (which is relatively clean) is at high temperature and contains recoverable heat energy. Figure 16.16 shows a schematic of a typical thermal incinerator system.

### 16.5.4.3 Catalytic Combustion

Catalytic combustion operates by passing a preheated contaminant-laden gas stream through a catalyst bed (usually a thinly coated platinum mesh mat, honeycomb, or other configuration designed to increase surface area),

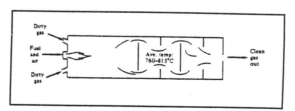

**Figure 16.15** Thermal incinerator. (*Source:* USPEA, *Control Techniques for Gases and Particulates,* 1968.)

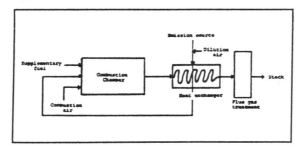

**Figure 16.16** Schematic of thermal incinerator system. (*Source:* Adaptation from Corbitt, 1990, *Environmental Engineering, p. 4.70.)

which promotes the oxidization reaction at lower temperatures (see Figure 16.17). The metal catalyst is used to initiate and promote combustion at much lower temperatures than those required for thermal combustion (metals in the platinum family are recognized for their ability to promote combustion at low temperature). Catalytic incineration may require 20 to 50 times less residence time than thermal incineration (see Table 16.3 for other advantages of catalytic incinerators over thermal incinerators). Catalytic incinerators normally operate at 700–900°F. At this reduced temperature range, a saving in fuel usage and cost is realized; however, this may be offset by the cost of the catalytic incinerator itself.

A schematic diagram of a catalytic incinerator is presented in Figure 16.18. A heat exchanger is an option for systems with heat transfer between two gas streams (recuperative heat exchange). The need for dilution air, combustion air, and/or flue gas treatment is based on site-specific conditions. Catalysts are subject to both physical and chemical deterioration, and their usefulness is suppressed by sulfur-containing compounds. For best performance, catalyst surfaces must be clean and active.

Catalytic incineration is used in a variety of industries to treat effluent

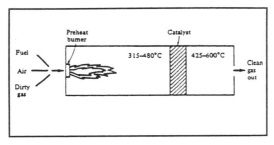

**Figure 16.17** A catalytic incinerator. (*Source:* USEPA, *Control Techniques for Gases and Particulates,* 1971.)

TABLE 16.3. Advantages of Catalytic over Thermal Incinerators.

| |
|---|
| (1) Catalytic incinerators have lower fuel requirements. |
| (2) Catalytic incinerators have lower operating temperatures. |
| (3) Catalytic incinerators have little or no insulation requirements. |
| (4) Catalytic incinerators have reduced fire hazards. |
| (5) Catalytic incinerators have reduced flashback problems. |

*Source:* Adaptation from Buonicore & Davis, 1992.

gases, including emissions from paint and enamel bake ovens, asphalt oxidation, coke ovens, formaldehyde manufacture, and varnish cooking.

## 16.6 REMOVAL OF GASEOUS POLLUTANTS: MOBILE SOURCES

Mobile sources of gaseous pollutants include locomotives, ships, airplanes, and automobiles. However, automobiles are by far the most important, both in terms of total emissions and in location of emissions relative to people. According to the *Twelfth Annual Report of the Council on Environmental Quality* (1982), transportation accounted for 55% of all major air pollutants emitted to the atmosphere in 1980. In 1986, almost 140 million passenger cars were registered in the United States—consuming more than 1 billion gallons of fuel. Total emissions from these vehicles included 58% of the nation's total carbon monoxide emissions, 38% of the lead, 34% of the nitrogen oxide, 27% of the VOCs, and 16% of the particulates (USEIA, 1988; USEPA, 1988).

Because of the high levels of pollutant emissions from the automobile, emission standards have become increasingly more stringent in the United States. Under the Clean Air Act of 1970, for example, motor vehicle emissions standards "forced" the development of new control technology to achieve compliance with standard emission levels.

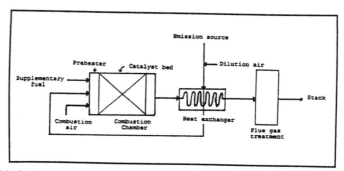

**Figure 16.18** Schematic of catalytic incinerator system. (*Source:* Adaptation from Corbitt, 1990, p. 4.71.)

Mobile source pollution problems can be solved by either of two means: replacing the internal combustion engine (e.g., with electrical power or mass transit) or by direct pollutant control systems. At the present time, replacement is not feasible, and technology in this regard is still in its infancy. Even if the technology were currently available for replacement, replacement would be inordinately difficult to undertake. Direct pollutant control systems (those that control emissions from the crankcase, carburetor, fuel tank, and exhaust) are what we rely on. We discuss these systems in the following section.

### 16.6.1 CONTROL OF CRANKCASE EMISSIONS

Crankcase emissions can be controlled by technology called positive crankcase ventilation (PCV). In this control technology, hydrocarbon blowby gases (gases that go past the piston rings into the crankcase) are recirculated to the combustion chamber for reburning (see Figure 16.19). The National Air Pollution Control Administration (1970) estimated that incorporation of the PCV system reduced crankcase hydrocarbon emissions to negligible levels.

### 16.6.2 CONTROL OF EVAPORATIVE EMISSIONS

Changes in ambient temperatures (diurnal loses), hot soak, and running losses result in evaporative emissions. Diurnal losses are caused by expansions of the air–fuel mixture in a partially filled fuel tank, which expels

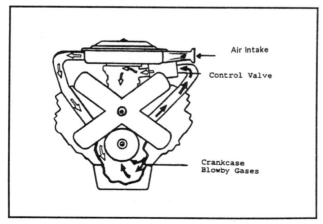

**Figure 16.19** Positive crankcase ventilation (PCV) of blowby gases. (*Source:* Adaptation from Godish 1997, p. 290.)

TABLE 16.4. **Types of Catalytic Converters.**

*Oxidizing catalytic converter:* works to accelerate the completion of the oxidation of CO and hydrocarbons so that CO is converted to carbon dioxide and water vapor. Platinum and palladium are used as catalysts. These converters can be poisoned by lead; thus, only unleaded gasoline should be used with this type.

*Reducing catalytic converter:* uses rhodium and ruthenium to accelerate the reduction of $NO_x$ to $N_2$.

*Three-way catalytic converter:* favored by U.S. automobile manufacturers because it enables them to meet compliance requirements of the Clean Air Act. Oxidizes hydrocarbons and carbon monoxide to carbon dioxide, while reducing $NO_x$ to $N_2$. Effective in controlling emissions and has the advantage of allowing the engine to operate at normal conditions where engine performance and efficiency are greatest.

*Source:* Adaptation from Demmler, 1977.

gasoline vapor into the atmosphere. Hot soak emissions occur after the engine is shut off as heat from the engine causes increased evaporation of fuel. Running (or operating) losses occur during driving as the fuel is heated by the road surface and when fuel is forced from the fuel tank while the vehicle is being operated and the fuel tank becomes hot. In 1971, the direct control measure instituted to control hydrocarbon emissions was the installation of a canister filled with activated charcoal that adsorbs hydrocarbon emissions. Adsorbed vapors are purged from charcoal into the engine during high-power operating conditions. In California, vapor control systems at service stations reduce potential refueling vapor losses (Perkins, 1974).

## 16.6.3 CATALYTIC CONVERTERS

Beginning in 1975, all new U.S. automobiles were required to be equipped with catalytic converters to meet the more restrictive tailpipe emission standards. Three types of catalytic converters are used for this purpose: oxidizing, reducing, and three-way. Table 16.4 lists characteristics of each of these catalytic converters.

## 16.7 SUMMARY

Regulatory agencies, consumers, environmentalists, and industries have, for some years now, participated in a square dance—participants shifting in and out of the center, forming ever-shifting partnerships, swinging in and out, around, past and in between the issues that concern our environment. Scientists assess environmental damage. Communities take action to effect change. Regulatory agencies create, monitor, and enforce the laws to control the problem issues, and industries walk the tightrope between profit and loss—between business life and death. Occasionally, one group or another forgets that everything we do is connected—that the actions of one group affect the rest—and that we are all in this together. Do you drive a car? Do you buy and use factory-made products? Then you've bought into industry. We cannot separate the segments cleanly in our society; like a foodweb in any ecosystem, what we are and do is connected to everything around us. We're part of the dance—and share in the responsibility.

## 16.8 REFERENCES

American Industrial Hygiene Association, *Air Pollution Manual: Control Equipment, Part II.* Detroit: AIHA, 1968.

Boubel, R. W., Fox, D. L., Turner, D. B., & Stern, A. C., *Fundamentals of Air Pollution.* New York: Academic Press, 1994.

Buonicore, A. J. & Davis, W. T. (eds.), *Air Pollution Engineering Manual.* New York: Van Nostrand Reinhold, 1992.

Cooper, C. D. & Alley, F. C., *Air Pollution Control: A Design Approach.* Prospect Heights, IL: Waveland Press, Inc., 1990.

Corbitt, R. A., *Standard Handbook of Environmental Engineering.* New York: McGraw-Hill, 1990.

Davis, M. L. & Cornwell, D. A., *Introduction to Environmental Engineering.* New York: McGraw-Hill, 1991.

Demmler, A. W., "Automotive Catalysis," *Auto Eng.,* 85(3): 29, 32 (1977).

Godish, T., *Air Quality,* 3rd ed. Boca Raton, FL: Lewis Publishers, 1997.

Hesketh, H. E., *Air Pollution Control: Traditional and Hazardous Pollutants.* Lancaster, PA: Technomic Publishing Co., Inc., 1991.

Heumann, W. L., *Industrial Air Pollution Control Systems.* New York: McGraw-Hill, 1997.

National Air Pollution Control Administration: *Control Techniques for Hydrocarbon and Organic Solvent Emissions for Stationary Sources,* Document B, Publ. AP–68, Washington, D.C., 1970.

Peavy, H. S., Rowe, D. R., & Tchobanglous, G., *Environmental Engineering.* New York: McGraw-Hill, 1985.

Perkins, H. C., *Air Pollution.* New York: McGraw-Hill, 1974.

*Twelfth Annual Report of the Council of Environmental Quality,* Washington, D.C., 1982.

USHEW, *Control Techniques for Particulate Air Pollutants,* Washington, D.C.: National Air Pollution Control Administration, 1969.

USEIA, *Annual Energy Review 1987,* Washington, D.C.: Energy Information Agency, Department of Energy, 1988.

USEPA, *Annual Report of the Environmental Protection Agency to the Congress of the United States in Compliance with Section 202(b)(4),* Public Law 90–148, Washington, D.C., 1971.

USEPA, *Air Pollution Engineering Manual,* 2nd ed., U.S. Environmental Protection Agency, AP–40, Research Triangle Park, NC, 1973.

USEPA, *Handbook—Control Technologies for Hazardous Air Pollutants,* U.S. Environmental Protection Agency, Center for Environmental Research Information, EPA 625/6–86/014, Cincinnati, OH, 1986.

USEPA, *National Air Pollutant Emission Estimates 1940–1986,* Environmental Protection Agency, Washington, D.C., 1988.

# CHAPTER 17

# Indoor Air Quality

*The quality of the air we breathe and the attendant consequences for human health are influenced by a variety of factors. These include hazardous material discharges indoors and outdoors, meteorological and ventilation conditions, and pollutant decay and removal processes. Over 80% of our time is spent in indoor environments so that the influence of building structures, surfaces, and ventilation are important considerations when evaluating air pollution exposures. (Wadden & Scheff, 1983, p. 1)*

## 17.1 INTRODUCTION

IN Chapter 11 microclimates were discussed, but we are all more closely influenced by certain kinds of microclimates than we realize. Microclimates we don't often think about are the indoor microclimates we spend 80% of our time in: the office and/or the home.

In fact, not much attention was given to indoor microclimates until after two events took place a few years ago. The first event had to do with Legionnaires' Disease and the second with sick building syndrome. These topics and a discussion of the causal factors leading to indoor air pollution are covered in this chapter.

## 17.2 LEGIONNAIRES' DISEASE

Ever since that infamous event occurred in Philadelphia in 1976 at the Belleview Stratford Hotel (during a convention of American Legion members), which included 182 cases and 29 deaths, the deadly bacterium *Legionella pneumophila* has become synonymous with the term *Legionnaires' Disease*. The deaths were attributed to bacteria in the hotel's cooling tower.

Organisms of the genus *Legionella* are ubiquitous in the environment and are found in natural fresh water and potable water, as well as in closed-circuit systems such as evaporative condensers, humidifiers, recreational whirlpools, air handling systems, and, of course, in cooling tower water.

The potential for the presence of *Legionella* bacteria is dependent on certain environmental factors: moisture, temperature (50°–140°F), oxygen, and a source of nourishment such as slime or algae.

Not all the ways in which Legionnaires' Disease can be spread are known to us at this time; however, we do know that it can be spread through the air. The Centers for Disease Control (CDC) states (in its *Questions and Answers on Legionnaires' Disease,* CDC No. 28L0343779) that there is no evidence that it is spread person-to-person.

Air-conditioning cooling towers and evaporative condensers have been the source of most outbreaks to date, and the bacterium is commonly found in both. Unfortunately, we do not know if this is an important means of spreading of Legionnaires' Disease because other outbreaks have occurred in buildings that did not have air conditioning.

Not all people are at risk of contracting Legionnaires' Disease. The people most at risk include

(1) Persons with lowered immunological capacity
(2) Cigarette smokers and those with a history of alcohol abuse
(3) Individuals exposed to high concentrations of *Legionella pneumophila*

Most commonly recognized as a form of pneumonia, the symptoms of Legionnaires' Disease usually become apparent two to ten days after known or presumed exposure to airborne Legionnaires' Disease bacteria. A sputum-free cough is common, but sputum production is sometimes associated with the disease. Within less than a day, the victim can experience rapidly rising fever and the onset of chills. Mental confusion, chest pain, abdominal pain, impaired kidney function, and diarrhea are associated manifestations of the disease. CDC estimates that about 25,000 people develop Legionnaires' Disease annually.

The obvious question is "how do we prevent or control Legionnaires' Disease?" The controls presently being used are targeted on cooling towers and air-handling units (condensate drain pans).

For cooling towers, procedures used to control bacterial growth vary somewhat on the various regions in a cooling tower system. However, control procedures usually include a good maintenance program, including repair/replacement of damaged components, routine cleaning, and sterilization.

In sterilization, a typical protocol calls for the use of chlorine in a residual solution at about 50 ppm (combined with a compatible detergent) to produce the desired sterilization effect. Ensuring that even somewhat inaccessible spaces are properly cleaned of slime and algae accumulations is essential.

Control measures for air-handling units' condensate drain pans typically involve keeping the pans clean and checked for proper drainage of fluid, which is essential in preventing stagnation and the buildup of slime, algae, and bacteria. A cleaning and sterilization program is required anytime algae or slime are found in the unit.

## 17.3 SICK BUILDING SYNDROME

The second event to catch the public's attention regarding the possibility of unhealthy indoor environments was actually spawned by the Legionnaires' event, by growing awareness of indoor air quality problems, and by other related incidents and complaints that followed. The term *sick building syndrome* (SBS) was coined by an international working group under the World Health Organization (WHO) in 1982.

The WHO working group studied the literature about indoor climate problems and found that some microclimates in buildings were characterized by a set of frequently appearing complaints and symptoms. WHO came up with five categories of symptoms or complaints reported by occupants who suffer from sick building syndrome. These categories are listed below:

(1) *Sensory irritation in eyes, nose, and throat:* pain, sensation of dryness, smarting feeling, stinging, irritation, hoarseness, voice problems

(2) *Neurological or general health symptoms:* headache, sluggishness, mental fatigue, reduced memory, reduced capability to concentrate, (dizziness, intoxication, nausea and vomiting, tiredness

(3) *Skin irritation:* pain, reddening, smarting or itching sensations, dry skin

(4) *Nonspecific hypersensitivity reactions:* runny nose and eyes, asthma-like symptoms among nonasthmatics, sounds from the respiratory system

(5) *Odor and taste symptoms:* changed sensitivity of olfactory or gustatory sense, unpleasant olfactory or gustatory perceptions

In the past, similar symptoms had been used to define other syndromes—the building disease, the building illness syndrome, building-related illness, or the tight-fitting office syndrome. In many cases, these appear synonymous with the sick building syndrome; thus, the WHO definition of SBS worked to combine these syndromes into one general definition or summary. A summary compiled by WHO (1982, 1984) and by Molhave (1986) of this combined definition includes the five categories of symptoms listed earlier along with the following:

(1) Irritation of mucous membranes in the eye, nose, and throat is among the most frequent symptoms.

(2) Other symptoms (e.g., from lower airways or from internal organs) should be infrequent.

(3) A large majority of occupants report symptoms.

(4) The symptoms appear especially frequent in one building or in part of it.

(5) No evident causality can be identified in relation either to exposures or to occupant sensitivity.

The WHO group suggested the possibility that the SBS symptoms have a common causality and mechanism (WHO, 1982). However, the existence of SBS is still a postulate because the descriptions of the symptoms in the literature are anecdotal and unsystematic (Molhave, 1992).

### 17.4 A COTTAGE INDUSTRY IS BORN

As a result of the Legionnaires' incident in Philadelphia, the WHO study, and significant amounts of media attention, sick building syndrome became a common term used by ordinary working people to describe their workplaces. An off-shoot of this media attention was a new cottage industry consisting of so-called "experts" in indoor air pollution who commenced selling their services to conduct sick building surveys. From about 1985 until the early 1990s, this new industry was booming. Since then, many of these new enterprises folded their operations because of a lack of business; the initial scare wore off.

During its heyday, the movement to solve the sick building syndrome problem resulted in some air pollutants being identified as culprits or potential culprits in causing SBS. One such category of air pollutants in nonindustrial environments was identified as volatile organic compounds (VOCs). In 1989, WHO classified (according to boiling point) the entire range of organic pollutants into four groups: (1) very volatile (gaseous compounds), (2) volatile organic compounds, (3) semivolatile organic compounds, and (4) organic compounds associated with particulate matter.

### 17.5 INDOOR AIR POLLUTION

Why is indoor air pollution a problem? As previously indicated, recognition that the indoor air environment may be a health problem is a relatively recent emergence. The most significant problems of indoor air quality are caused by the impact of cigarette smoking; stove and oven operation; and emanations from certain types of particleboard, cement, and other building materials (Wadden & Scheff, 1983).

The significance of the indoor air quality problem became apparent, not only because of the Legionnaires' incident of 1976 and the WHO study of 1982, but also because of another factor that came to the forefront in the mid-1970s—the need to conserve energy. In the early 1970s, when hundreds of thousands of people were waiting in line to obtain gasoline for their automobiles, driving home the need to conserve energy supplies was not difficult.

The resulting impact of energy conservation on inside environments has been substantial, especially regarding building modifications made to decrease ventilation rates and new construction practices incorporated to ensure "tight" buildings to minimize infiltration of outdoor air.

There is some irony in this development, of course. While we do have a need to ensure proper building design, construction, and ventilation guidelines to avoid exposing inhabitants to unhealthy environments, what really resulted in this mad dash to reduce ventilation rates and "tighten" buildings from infiltration was a trade-off—energy economics versus air quality.

## 17.6  COMMON INDOOR AIR POLLUTANTS

This section takes a brief source-by-source look at the most common indoor air pollutants, their potential health effects, and ways to reduce their levels in the home or other indoor environments.

### 17.6.1 RADON

Radon is a nontoxic, colorless, odorless noble gas produced in the decay of radium-226 and found everywhere at very low levels. Radon is ubiquitously present in the soil and air near the surface of the Earth. As radon undergoes radioactive decay, it releases an alpha particle, gamma ray, and progeny that quickly decay to release alpha and beta particles and gamma rays. Because radon progeny are electrically charged, they readily attach to particles, producing a radioactive aerosol. When radon becomes trapped in buildings and concentrations build up in indoor air, exposure to radon becomes of concern because aerosol radon-contaminated particles may be inhaled and deposited in the bifurcations of respiratory airways. Irradiation of tissue at these sites poses a significant risk of lung cancer (depending on exposure dose).

The most common way in which radon enters a house is through the soil or rock upon which the house is built. The most common source of indoor radon is uranium, which is common to many soils and rocks. As uranium breaks down, it releases radon gas, which breaks down into radon decay products or progeny (commonly called radon daughters). Radon gas is transported into buildings by pressure-induced convective air flows. Other sources of radon include well water and masonry materials.

Radon levels in a house vary in response to temperature-dependent and wind-dependent pressure differentials and to changes in barometric pressures. When the base of a house is under significant negative pressure, radon transport is enhanced.

Studies by the EPA (1987, 1988) indicate that as many as 10% of all American homes (about 9 million homes) may have elevated levels of radon, and the percentage may be higher in geographic areas with certain soils and bedrock formations.

According to the USEPA's booklets, *A Citizen's Guide to Radon, Radon Reduction Methods: A Homeowner's Guide,* and *Radon Measurement Proficiency Report* (for each state), exposure to radon in the home can be reduced by the following steps:

(1) Measure levels of radon in the home.
(2) Contact the state radiation protection office for information on the availability of detection devices or services.
(3) Refer to EPA guidelines in deciding whether (and how quickly) to take action based on test results.
(4) Learn about control measures.
(5) Take precautions not to draw larger amounts of radon into the house.
(6) Select a qualified contractor to draw up and implement a radon mitigation plan.
(7) Stop smoking and discourage smoking in your home.
(8) Treat radon-contaminated well water by aerating or filtering it through granulated activated charcoal.

## 17.6.2 ENVIRONMENTAL TOBACCO SMOKE

The use of tobacco products by approximately 45 million smokers in the United States results in significant indoor contamination from combustion by-products that pose significant exposures to millions of others who do not smoke, but who must breathe contaminated indoor air. Composed of sidestream smoke (smoke that comes from the burning end of a cigarette) and smoke exhaled by the smoker, tobacco smoke contains a complex mixture of over 4700 compounds, including both gases and particulates.

According to reports issued in 1986 by the Surgeon General and the National Academy of Sciences, environmental tobacco smoke is a cause of disease (including lung cancer) in both smokers and healthy nonsmokers. Environmental tobacco smoke may also increase the lung cancer risk associated with exposures to radon.

The following steps can reduce exposure to environmental tobacco smoke in the office or home:

(1) Give up smoking and discourage smoking in your home and place of work, or require smokers to smoke outdoors.

(2) Ventilate your home or office, which helps to reduce—but not eliminate—exposure.

### 17.6.3 BIOLOGICAL CONTAMINANTS

A variety of biological contaminants can cause significant illness and health risks. These include mold and mildew, animal dander, cat saliva, mites, cockroaches, and pollen, as well as airborne exposure to viruses that cause colds and influenza and exposure to the bacteria that causes Legionnaires' Disease and tuberculosis (TB).

The following steps can reduce exposure to biological contaminants in the home and office.

(1) Install and use exhaust fans that are vented to the outdoors in kitchens and bathrooms, and vent clothes dryers outdoors.

(2) Ventilate the attic and crawl spaces to prevent moisture build-up.

(3) Keep water trays in cool mist or ultrasonic humidifiers clean and filled with fresh distilled water daily.

(4) Water-damaged carpets and building materials should be thoroughly dried and cleaned within 24 hours.

(5) Maintain good housekeeping practices both in the home and office.

### 17.6.4 COMBUSTION BY-PRODUCTS

Combustion by-products are released into indoor air from a variety of sources, including unvented kerosene and gas space heaters, woodstoves, fireplaces, gas stoves, and hot water heaters. The major pollutants released from these sources are carbon monoxide, nitrogen dioxide, and particulates.

The following steps can reduce exposure to combustion products in the home and office:

(1) Fuel-burning unvented space heaters should only be operated using great care and special safety precautions.

(2) Install and use exhaust fans over gas cooking stoves and ranges, and keep the burners properly adjusted.

(3) Furnaces, flues, and chimneys should be inspected annually, and any needed repairs should be made promptly.

(4) Woodstove emissions should be kept to a minimum.

### 17.6.5 HOUSEHOLD PRODUCTS

A large variety of organic compounds is widely used in household products because of their useful characteristics such as the ability to dissolve substances and evaporate quickly. Cleaning, disinfecting, cosmetic, degreasing, and hobby products all contain organic solvents, as do paints, varnishes, and waxes. All of these products can release organic compounds while you use them, as well as when they are stored.

The following steps can reduce exposure to household organic compounds:

- Always follow label instructions carefully.
- Dispose of partially full chemical containers safely.
- Limit the amount you buy to what you can immediately use.

### 17.6.6 PESTICIDES

Pesticides represent a special case of chemical contamination of buildings, where the EPA estimates 80–90% of most people's pesticide exposure in the air occurs. These products are extremely dangerous if not used properly.

The following steps can reduce exposure to pesticides in the home:

- Read the label and follow directions.
- Use pesticides only in well-ventilated areas.
- Dispose of unwanted pesticides safely.

### 17.6.7 ASBESTOS

Asbestos became a major indoor air quality concern in the United States in the late 1970s. Asbestos is a mineral fiber once commonly used in a variety of building materials and now identified as having the potential (when reduced to fibers) to cause cancer in humans.

The following steps can reduce exposure to asbestos in the home or office:

- Do not cut, rip, or sand asbestos-containing materials.
- When you need to remove or clean up asbestos, use a professionally trained contractor.

## 17.7 AIR QUALITY CONTROL MEASURES

Three basic strategies are presently being used to improve indoor air quality: source control (usually the best method), ventilation improvements, and the use of mechanical devices such as air cleaners. An excellent resource on residential air cleaners is the EPA's *Air and Radiation* booklet (1990).

## 17.8 SUMMARY

Indoor air quality was a blind spot in air quality control for many years. Increased awareness of air quality issues began with Legionnaires' Disease and is now brought to the forefront again by public awareness of the dangers of second-hand smoke. Most of us now live and work indoors most of the time; we are no longer an agrarian society. Instead, in our modern industrial society, our lives are bound by walls, and the air we breathe is often as processed as the clothes we wear and the food we eat.

## 17.9 REFERENCES

Molhave, L., "Indoor air quality in relation to sensory irritation due to volatile organic compounds," *ASHRAE Trans.* 92(1):306–316, Publication #2954, 1986.

USEPA, *A Citizen's Guide to Radon*, 1986.

USEPA, Indoor Air Facts No. 1, *EPA and Indoor Air Quality*, 1987.

USEPA, *The Inside Story—A Guide to Indoor Air Quality*, 1988.

USEPA, Residential Air Cleaners Indoor Air Facts No. 7, *Air and Radiation*, 1990.

Wadden, R. A. & Scheff, P. A., *Indoor Air Pollution: Characteristics, Predications, and Control.* New York: John Wiley, 1983.

World Health Organization (WHO). *Indoor Air Pollutants, Exposure and Health Effects Assessment.* Euro-Reports and Studies No. 78: Working Group Report. Copenhagen: WHO Regional Office, 1982.

World Health Organization (WHO). *Indoor Air Quality Research.* Euro-Reports and Studies No. 103. Copenhagen: WHO Regional Office, 1984.

# Afterword

**O**NE hundred years ago, modern industry was in its infancy. Consider the advances made in science and technology in the last 100 years, and think of what may lie ahead in the next 100 years—what could occur is staggering. Information, knowledge, and science are cumulative. They build on what came before, and the broader the foundation of knowledge and information, the broader the base of science and higher we can build and climb.

Mankind is on the brink of a new age of exploration. This age may be delayed for a while—it may not seriously begin within our lifetime—but the approach has begun. On our own planet, our ocean depths still remain largely a mystery, and space beckons us outward to the stars.

These two environments present a common concern and a serious problem to be solved in their exploration: the lack of air. However people encompass themselves in the trappings of technology or surround themselves with failsafes and backups, under the ocean and out in space, a serious failure of technology means we die.

Our science, our technology, our information provides us with the ability to create and to expand our vision and our horizons. We insulate and isolate ourselves from the natural world and increasingly barricade ourselves from nature and our environment with technology, but beneath the industrial, social, and psychological structures we build for ourselves, we are still human and vulnerable to the world around us.

As is our world vulnerable to us. The science of nature and of our natural world includes amazements, mysteries, wonders to explore, and processes and evolutions we could never have imagined on our own, as well as some cold equations. Without clear air, clean water, and a clean environment, we die.

As the world's population grows, as we must stretch our natural resources to the limits, as we build deeper into the Earth and higher on its surface, and as we expand our horizons, the science and technology of the environment and of our natural world become increasingly important. The pressures and problems facing environmental science will not lessen. They can only increase.

# Index